Wireless Optical
Communication Systems

WIRELESS OPTICAL
COMMUNICATION SYSTEMS

STEVE HRANILOVIC
Assistant Professor
Department of Electrical and Computer Engineering
McMaster University
Hamilton, Ontario, Canada

 Springer

Steve Hranilovic
McMaster University
Dept. of Electrical & Computer Engineering
1280 Main Street W.
Hamilton, ONT L8S 4K1
Canada
hranilovic@ece.mcmaster.ca

Hranilovic, Steve, 1973-
   Wireless optical communication systems / Steve Hranilovic
      p. cm.
   Includes bibliographical references and index.
   ISBN 0-387-22785-7 (e-Book)
   1. Optical communications. 2. Wireless communication systems. I. Title.

TK5103.59.H73 2004
621.382'7--dc22
                                        2004051193
ISBN 978-1-4419-1982-3   e-ISBN 978-0-387-22785-6

Printed on acid-free paper.

Printed in the United States of America.

9 8 7 6 5 4 3 2 1

springeronline.com

To Annmarie

Contents

Preface

The use of optical free-space emissions to provide indoor wireless communications has been studied extensively since the pioneering work of Gfeller and Bapst in 1979 [1]. These studies have been invariably interdisciplinary involving such far flung areas such as optics design, indoor propagation studies, electronics design, communications systems design among others. The focus of this text is on the design of communications systems for indoor wireless optical channels. Signalling techniques developed for wired fibre optic networks are seldom efficient since they do not consider the bandwidth restricted nature of the wireless optical channel. Additionally, the elegant design methodologies developed for electrical channels are not directly applicable due to the amplitude constraints of the optical intensity channel. This text is devoted to presenting optical intensity signalling techniques which are *spectrally efficient*, i.e., techniques which exploit careful pulse design or spatial degrees of freedom to improve data rates on wireless optical channels.

The material presented here is complementary to both the comprehensive work of Barry [2] and to the later book by Otte *et al.* [3] which focused primarily on the design of the optical and electronic sub-systems for indoor wireless optical links. The signalling studies performed in these works focused primarily on the analysis of popular signalling techniques for optical intensity channels and on the use of conventional electrical modulation techniques with some minor modifications (e.g., the addition of a bias). In this book, the design of spectrally efficient signalling for wireless optical intensity channels is approached in a fundamental manner. The goal is to extend the wealth of modem design practices from electrical channels to optical intensity domain. Here we discuss important topics such as the vector representation of optical intensity signals, the design and capacity of signalling sets as well as the use of multiple transmitter and receiver elements to improve spectral efficiency.

Although this book is based on my doctoral [4] and Masters [5] theses, it differs substantially from both in several ways. Chapters 2 and 3 are completely re-written and expanded to include a more tutorial exposition of the basic issues involved in signalling on wireless optical channels. Chapters 4-6, which develop the connection between electrical signalling design and optical intensity channels, are significantly re-written in more familiar language to allow them to be more accessible. Chapters 7 and 8 are improved through the addition of a fundamental analysis of MIMO optical channels and the increase in capacity which arise due to spatial multiplexing in the presence of spatial bandwidth constraints. Significant background material has been added on the physical aspects of wireless optical channels including optoelectronic components and propagation characteristics to serve as an introduction to communications specialists. Additionally, fundamental communication concepts are briefly reviewed in order to make the signalling design sections accessible to experimentalists and applied practitioners.

Finally, there have been a great number of individuals who have influenced the writing of this book and deserve my thanks. I am very grateful to my doctoral thesis advisor Professor Frank R. Kschischang who's passion for research and discovery have inspired me. Additionally, I would like to thank Professors David A. Johns and Khoman Phang for introducing me to the area and for fostering my early explorations in wireless optical communications. I am also indebted to a number of friends and colleagues who have contributed through many useful conversations, among them are : Warren Gross, Yongyi Mao, Andrew Eckford, Sujit Sen, Tooraj Esmailian, Terence Chan, Masoud Ardakani and Aaron Meyers.

Foremost, I would like to thank my wife Annmarie for her patience, understanding and for her support.

STEVE HRANILOVIC

PART I

INTRODUCTION

Chapter 1

INTRODUCTION

In recent years, there has been a migration of computing power from the desktop to portable, mobile formats. Devices such as digital still and video cameras, portable digital assistants and laptop computers offer users the ability to process and capture vast quantities of data. Although convenient, the interchange of data between such devices remains a challenge due to their small size, portability and low cost. High performance links are necessary to allow data exchange from these portable devices to established computing infrastructure such as backbone networks, data storage devices and user interface peripherals. Also, the ability to form *ad hoc* networks between portable devices remains an attractive application. The communication links required can be categorized as short-range data interchange links and longer-range wireless networking applications.

One possible solution to the data interchange link is the use of a direct electrical connection between portable devices and a host. This electrical connection is made via a cable and connectors on both ends or by some other direct connection method. The connectors can be expensive due to the small size of the portable device. In addition, these connectors are prone to wear and break with repeated use. The physical pin-out of the link is fixed and incompatibility among various vendors solutions may exist. Also, the need to carry the physical medium for communication makes this solution inconvenient for the user.

Wireless radio frequency (RF) solutions alleviate most of the disadvantages of a fixed electrical connection. RF wireless solutions allow for indoor and short distance links to be established without any physical connection. However, these solutions remain relatively expensive and have low to medium data rates. Some popular "low cost" RF links over distances of approximately 10 m provide data rates of up to 1 Mbps in the 2.4 GHz band for a cost near US$5 per module. Indoor IEEE 802.11 [6] links have also gained significant popularity

Table 1.1.   Comparison of wireless optical and radio channels.

Property	Wireless Optical	Radio
Cost	$	$$
RF circuit design	No	Yes
Bandwidth Regulated	No	Yes
Data Rates	100's Mbps	10's Mbps
Security	High	Low
Passes through walls ?	No	Yes

and provide data rates of approximately 50 Mbps. Radio frequency wireless links require that spectrum licensing fees are paid to federal regulatory bodies and that emissions are contained within strict spectral masks. These frequency allocations are determined by local authorities and may vary from country to country, making a standard interface difficult. In addition, the broadcast nature of the RF channel allows for mobile connectivity but creates problems with interference between devices communicating to a host in close proximity. Containment of electromagnetic energy at RF frequencies is difficult and if improperly done can impede system performance.

This book considers the use of wireless optical links as another solution to the short-range interchange and longer-range networking links. Table 1.1 presents a comparison of some features of RF and wireless optical links. Present day wireless optical links can transmit at 4 Mbps over short distances using opto-electronic devices which cost approximately US$1 [7]. However, much high rates approaching 1 Gpbs have been investigated in some experimental links. Wireless optical links transmit information by employing an optoelectronic light modulator, typically a light-emitting diode (LED). The task of up- and down-conversion from baseband frequencies to transmission frequencies is accomplished without the use of high-frequency RF circuit design techniques, but is accomplished with inexpensive LEDs and photodiodes. Since the electromagnetic spectrum is not licensed in the optical band, spectrum licensing fees are avoided, further reducing system cost. Optical radiation in the infrared or visible range is easily contained by opaque boundaries. As a result, interference between adjacent devices can be minimized easily and economically. Although this contributes to the security of wireless optical links and reduces interference it also impacts rather stringently on the mobility of such devices. For example, it is not possible for a wireless optical equipped personal digital assistant to communicate if it is stored in a brief case. Wireless optical links are also suited to portable devices since small surface mount light emitting and light detecting components are available in high volumes at relatively low cost.

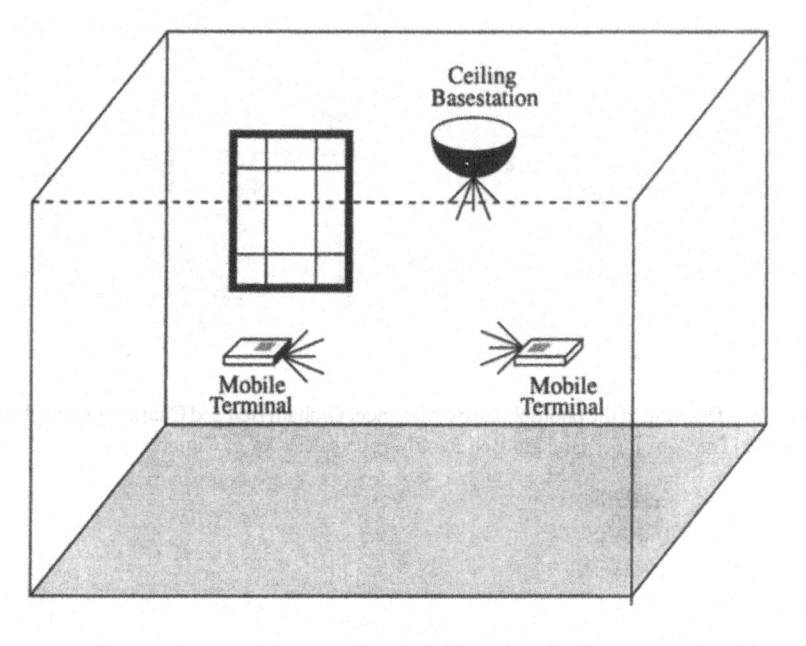

Figure 1.1. An indoor wireless optical communication system.

Figure 1.1 presents a diagram of a typical indoor wireless optical communications scenario. Mobile terminals are allowed to roam inside of a room and require that links be established with a ceiling basestation as well as with other mobile terminals. In some links the radiant optical power is directed toward the receiver, while in others the transmitted signal is allowed to bounce diffusely off surfaces in the room. Ambient light sources are the main source of noise in the channel and must be considered in system design. However, the available bandwidth in some directed wireless optical links can be large and allows for the transmission of large amounts of information, especially in short range applications.

Indoor wireless optical communication systems are envisioned here as a complimentary rather than a replacement technology to RF links. Whereas, RF links allow for greater mobility wireless optical links excel at short-range, high-speed communications such as in device interconnection or board-to-board interconnect.

1.1 A Brief History of Wireless Optical Communications

The use of optical emissions to transmit information has been used since antiquity. Homer, in the *Iliad*, discusses the use of optical signals to transmit a message regarding the Grecian siege of Troy in approximately 1200 BC. Fire beacons were lit between mountain tops in order to transmit the message

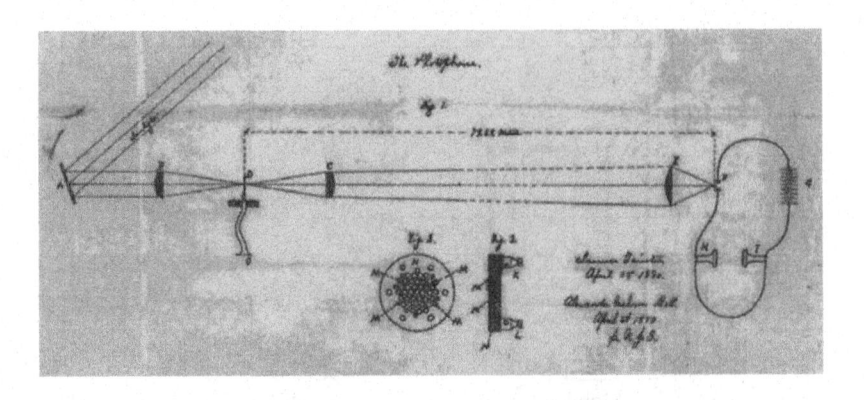

Figure 1.2. Drawing of the photophone by Alexander Graham Bell and Charles Sumner Tainter, April 1880 [The Alexander Graham Bell Family Papers, Library of Congress].

over great distances. Although the communication system is able to only ever transmit a single bit of information, this was by far the fastest means to transmit information of important events over long distances.

In early 1790's, Claude Chappe invented the optical telegraph which was able to send messages over distances by changing the orientation of signalling "arms" on a large tower. A code book of orientations of the signalling arms was developed to encode letters of the alphabet, numerals, common words and control signals. Messages could be sent over distances of hundreds of kilometers in a matter of minutes [8].

One of the earliest wireless optical communication devices using electronic detectors was the *photophone* invented by A. G. Bell and C. S. Tainter and patented on December 14, 1880 (U.S. patent 235,496). Figure 1.2 presents a drawing made by the inventors outlining their system. The system is designed to transmit a operator's voice over a distance by modulating reflected light from the sun on a foil diaphragm. The receiver consisted of a selenium crystal which converted the optical signal into an electrical current. With this setup, they were able to transmit an audible signal a distance of 213 m [9].

The modern era of indoor wireless optical communications was initiated in 1979 by F.R. Gfeller and U. Bapst by suggesting the use of diffuse emissions in the infrared band for indoor communications [1]. Since that time, much work has been done in characterizing indoor channels, designing receiver and transmitter optics and electronics, developing novel channel topologies as well as in the area of communications system design. Throughout this book, previous work on a wide range of topics in wireless optical system will be surveyed.

1.2   Overview

The study of wireless optical systems is multidisciplinary involving a wide range of areas including: optical design, optoelectronics, electronics design, channel modelling, communications and information theory, modulation and equalization, wireless optical network architectures among many others.

This book focuses on the issues of signalling design and information theory for wireless optical intensity channels. This book differs from Barry's comprehensive work *Wireless Infrared Communications* [2] and the text by Otte *et al. Low-Power Wireless Infrared Communications* by focusing exclusively on the design of modulation and coding for single element and multi-element wireless optical links. This work is complimentary and focuses on the design of signalling and communication algorithms for wireless optical intensity channels.

The design of a communication algorithms for any channel first requires knowledge of the channel characteristics. Chapter 2 overviews the basic operation of optoelectronic devices and the amplitude constraints that they introduce. Eye and skin safety, channel propagation characteristics, noise and a variety of channel topologies are described.

Most signalling techniques for wireless optical channels are adapted from wired optical channels. Conventional signalling design for the electrical channel cannot be applied to the wireless optical intensity channel due to the channel constraints. A majority of signalling schemes for optical intensity channels deal with binary-level on-off keying or PPM. Although power efficient, their spectral efficiency is poor. Chapter 3 overviews basic concepts in communications system design such as vector channel model, signal space, bandwidth as well a presenting an analysis of some popular binary and multi-level modulation schemes.

Part II of this book describes techniques for the design and analysis of spectrally efficient signalling techniques for wireless optical channels. This work generalizes previous work in optical intensity channels in a number of important ways. In Chapter 4, a signal space model is defined which represents the amplitude constraints and the cost geometrically. In this manner, all time-disjoint signalling schemes for the optical intensity channel can be treated in a common framework, not only rectangular pulse sets.

Having represented the set of transmittable signals in signal space, Chapter 5 defines lattice codes for optical intensity channels. The gain of these codes over a baseline is shown to factor into coding and shaping gains. Unlike previous work, the signalling schemes are not confined to use rectangular pulses. Additionally, a more accurate bandwidth measure is adopted which allows for the effect of shaping on the spectral characteristics to be represented as an effective dimension. The resulting example lattice codes which are defined show that on an idealized point-to-point link significant rate gains can be had by using spectrally efficient pulse shapes.

Chapter 6 presents bounds on the capacity of optical intensity signalling sets subject to an average optical power constraint and a bandwidth constraint. Although the capacity of Poisson photon counting channels has been extensively investigated, the wireless optical channel is Gaussian noise limited and pulse sets are not restricted to be rectangular. The specific bounds on the channel capacity of wireless optical channels exist for the case of PPM signalling and multiple-subcarrier modulation. The bounds presented in this work generalize these previous results and allow for the direct comparison of convention rectangular modulation with more spectrally efficient schemes. The bounds are shown to converge at high optical signal-to-noise ratios. Applied to several examples, the bounds illustrate that spectrally efficient signalling is necessary to maximize transmit rate at high SNR.

The spectral efficiency and reliability of wireless optical channels can also be improved by using multiple transmitter and receiver elements. Part III considers the modelling and signalling problem of multi-element links. Chapter 7 discusses the use of multiple transmit and receive elements to improve the efficiency of wireless optical links and presents a discussion on the challenges which are faced in signalling design.The pixelated wireless optical channel is defined as a multi-element link which improves the spectral efficiency of links unlike previous multi-element links, such as quasi-diffuse links and angle diversity schemes,. Although chip-to-chip, inter-board and holographic storage systems exploit spatial diversity for gains in data rate, the pixelated wireless optical channel does not rely on tight spatial alignment or use a pixel-matched assumption. Chapter 8 presents an experimental multi-element link in order to develop a channel model based on measurements. Using this channel model pixel-matched and pixelated optical spatial modulation techniques are compared.

Finally, Chapter 9 presents concluding remarks and directions for further study.

Chapter 2

WIRELESS OPTICAL INTENSITY CHANNELS

Communication systems transmit information from a transmitter to a receiver through the construction of a time-varying physical quantity or a *signal*. A familiar example of such a system is a wired electronic communications system in which information is conveyed from the transmitter by sending an electrical current or voltage signal through a conductor to a receiver circuit. Another example is wireless radio frequency (RF) communications in which a transmitter varies the amplitude, phase and frequency of an electromagnetic carrier which is detected by a receive antenna and electronics.

In each of these communications systems, the transmitted signal is corrupted by deterministic and random distortions due to the environment. For example, wired electrical communication systems are often corrupted by random thermal as well as shot noise and are often frequency selective. These distortions due to external factors are together referred to as the response of a communications *channel* between the transmitter and receiver. For the purposes of system design, the communications channel is often represented by a mathematical *model* which is realistic to the physical channel. The goal of communication system design is to develop signalling techniques which are able to transmit data reliably and at high rates over these distorting channels.

In order to proceed with the design of signalling for wireless optical channels a basic knowledge of the channel characteristics is required. This chapter presents a high-level overview of the characteristics and constraints of wireless optical links. Eye and skin safety requirements as well as amplitude constraints of wireless optical channels are discussed. These constraints are fundamental to wireless optical intensity channels and do not permit the direct application of conventional RF signalling techniques. The propagation characteristics of optical radiation in indoor environments is also presented and contrasted to RF channels. The choice and operation of typical optoelectronics used in wire-

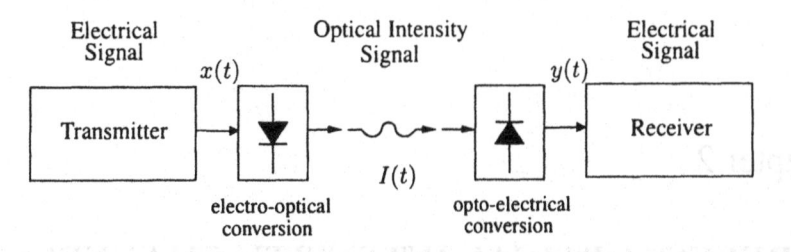

Figure 2.1.   Block Diagram of an optical intensity, direct detection communications channel.

less optical links is also briefly surveyed. Various noise sources present in the wireless optical link are also discussed to determine which are dominant. The chapter concludes with a comparison of popular channel topologies and a summary of the typical parameters of a practical short-range wireless optical channel.

2.1   Wireless Optical Intensity Channels

Wireless optical channels differ in several key ways from conventional communications channels treated extensively in literature. This section describes the physical basis for the various amplitude and power constraints as well as propagation characteristics in indoor environments.

2.1.1   Basic Channel Structure

Most present-day optical channels are termed intensity modulated, direct-detection channels. Figure 2.1 presents a schematic of a simplified free-space intensity modulated, direct-detection optical link.

The *optical intensity* of a source is defined as the optical power emitted per solid angle in units of Watts per steradian [10]. Wireless optical links transmit information by modulating the instantaneous optical intensity, $I(t)$, in response to an input electrical current signal $x(t)$. The information sent on this channel is not contained in the amplitude, phase or frequency of the transmitted optical waveform, but rather in the intensity of the transmitted signal. Present day optoelectronics cannot operate directly on the frequency or phase of the 10^{14} Hz range optical signal. This electro-optical conversion process is termed *optical intensity modulation* and is usually accomplished by a light-emitting diode (LED) or laser diode (LD) operating in the 850-950 nm wavelength band [11]. The electrical characteristics of the light emitter can be modelled as a diode, as shown in the figure. Section 2.2.1 describes the operation of LEDs and LDs in greater detail.

The opto-electrical conversion is typically performed by a silicon photodiode. The photodiode detector is said to perform *direct-detection* of the incident optical intensity signal since it produces an output electrical photocurrent, $y(t)$, nearly proportional to the received irradiance at the photodiode, in units of Watts per unit area [10]. Electrically, the detector is a reversed biased diode, as illustrated in Figure 2.1. Thus, the photodiode detector produces an output electrical current which is a measure of the optical power impinging on the device. The photodiode detector is often termed a *square law device* since the device can also be modelled as squaring the amplitude of the incoming electromagnetic signal and integrating over time to find the intensity. Section 2.2.2 describes the operation of p-i-n and avalanche type photodiodes and discusses their application to wireless optical channels.

The underlying structure of the channel, which allows for the modulation and detection of optical intensities only, places constraints on the class of signals which may be transmitted. The information bearing intensity signal which is transmitted must remain non-negative for all time since the transmitted power can physically never be negative, i.e.,

$$(\forall t \in \mathbb{R}) \; I(t) \geq 0. \tag{2.1}$$

Thus, the physics of the link imposes the fundamental constraint on signalling design that the transmitted signals remain non-negative for all time. In Chapters 4–6 this non-negativity constraint is taken into account explicitly in developing a framework for the design and analysis of modulation for optical intensity channels.

2.1.2   Eye and Skin Safety

Safety considerations must be taken into account when designing a wireless optical link. Since the energy is propagated in a free-space channel, the impact of this radiation on human safety must be considered.

There are a number of international standards bodies which provide guidelines on LED and laser emissions namely: the International Electrotechnical Commission (IEC) (IEC60825-1), American National Standards Institute (ANSI) (ANSI Z136.1), European Committee for Electrotechnical Standardization (CENELEC) among others. In this section, we will consider the IEC standard [12] which has been widely adopted. This standard classifies the main exposure limits of optical sources. Table 2.1 includes a list of the primary classes under which an optical radiator can fall. Class 1 operation is most desirable for a wireless optical system since emissions from products are safe under all circumstances. Under these conditions, no warning labels need to be applied and the device can be used without special safety precautions. This is important since these optical links are destined to be inexpensive, portable and convenient for the user. An extension to Class 1, termed Class 1M, refers to sources which are safe under normal operation but which may be hazardous if viewed with optical instruments [13]. Longer distance free-space links often operate in class 3B mode, and are used for high data rate transmission over moderate distances

Table 2.1. Interpretation of IEC safety classification for optical sources.*

Safety Class	Interpretation
Class 1	Safe under reasonably foreseeable conditions of operation.
Class 2	Eye protection afforded by aversion responses including blink reflex (for visible sources only λ=400–700 nm).
Class 3A	Safe for viewing with unaided eye. Direct intra-beam viewing with optical aids may be hazardous.
Class 3B	Direct intra-beam viewing is always hazardous. Viewing diffuse reflections is normally safe.

* Based on [12].

(40 m in [14]). The safety of these systems is maintained by locating optical beams on rooftops or on towers to prevent inadvertent interruption [15]. On some longer range links, even though the laser emitter is Class 3B, the system can still be considered Class 1M if appropriate optics are employed to spread the beam over a wide enough angle.

The critical parameter which determines whether a source falls into a given class depends on the application. The allowable exposure limit (AEL) depends on the wavelength of the optical source, the geometry of the emitter and the intensity of the source. In general, constraints are placed on both the peak and average optical power emitted by a source. For most practical high frequency modulated sources, the average transmitted power of modulation scheme is more restrictive than the peak power limitation and sets the AEL for a given geometry and wavelength [12]. At modulation frequencies greater than about 24 kHz, the AEL can be calculated based on average output power of the source [11].

The choice of which optical wavelength to use for the wireless optical link also impacts the AEL. Table 2.2 presents the limits for the average transmitted optical power for the IEC classes listed in Table 2.1 at four different wavelengths. The allowable average optical power is calculated assuming that the source is a point emitter, in which the radiation is emitted from a small aperture and diverges slowly as is the case in laser diodes. Wavelengths in the 650 nm range are visible red light emitters. There is a natural aversion response to high intensity sources in the visible band which is not present in the longer wavelength infrared band. The visible band has been used rarely in wireless optical communication applications due to the high background ambient light noise present in the channel. However, there has been some development of visible band wireless optical communications for low-rate signalling [16, 17]. Infrared wavelengths are typically used in optical networks. The wavelengths

Table 2.2. Point source safety classification based on allowable average optical power output for a variety of optical wavelengths.[*]

Safety Class	650 nm visible	880 nm infrared	1310 nm infrared	1550 nm infrared
Class 1	< 0.2 mW	< 0.5 mW	< 8.8 mW	< 10 mW
Class 2	0.2–1 mW	n/a	n/a	n/a
Class 3A	1–5 mW	0.5–2.5 mW	8.8–45 mW	10–50 mW
Class 3B	5–500 mW	2.5–500 mW	45–500 mW	50–500 mW

[*] Based on [15, 12].

λ=880 nm, 1310 nm and 1550 nm correspond to the loss minima in typical silica fibre systems, at which wavelengths optoelectronics are commercially available [18]. The trend apparent in Table 2.2 is that for class 1 operation the allowable average optical power increases as does the optical wavelength. This would suggest that the "far" infrared wavelengths above 1 μm are best suited to wireless optical links due to their higher optical power budget for class 1 operation. In this example, at least 20 times more optical power can be emitted with a 1550 nm source than with a 880 nm source. The difficulty in using this band is the cost associated with these far infrared devices. Photodiodes for far infrared bands are made from III-V semiconductor compounds while photodiodes for the 880 nm band are manufactured in low cost silicon technologies. Also, far-infrared components typically have smaller relative surface areas than their silicon near-infrared counterparts making the optical coupling design more challenging. As a result, the 880 nm "near" infrared optical band is typically used for inexpensive wireless optical links.

The power levels listed in Table 2.2 are pessimistic when applied to light sources which emit less concentrated beams of light, such as light emitting diodes. Indeed, more recent IEC and ANSI standards have recognized this fact and relaxed the optical power constraint for extended sources such as LEDs. However, the trends present in the table still hold. For a diode with λ=880 nm, a diameter of 1 mm and emitting light through a cone of angle 30°, the allowable average power for class 1 operation is 28 mW [11]. However, the allowed average optical power for class 1 operation still increases with wavelength. Section 2.2.1 discusses the tradeoff between the use of lasers or light emitting diodes as light emitters.

Eye safety considerations limit the average optical power which can be transmitted. This is another fundamental limit on the performance of free-space optical links.

Therefore, the constraint on any signalling scheme constructed for wireless optical links is that the *average* optical power is limited. As a result, the average amplitude (i.e., the normalized average optical power),

$$\lim_{T \to \infty} \frac{1}{2T} \int_{-T}^{T} I(t) \, dt \leq P, \tag{2.2}$$

for some fixed value P which satisfies safety regulations. This is in marked contrast to conventional electrical channels in which the constraint is on the averaged squared amplitude of the transmitted signal.

2.1.3   Channel Propagation Properties

As is the case in radio frequency transmission systems, multipath propagation effects are important for wireless optical networks. The power launched from the transmitter may take many reflected and refracted paths before arriving at the receiver. In radio systems, the sum of the transmitted signal and its images at the receive antenna cause spectral nulls in the transmission characteristic. These nulls are located at frequencies where the phase shift between the paths causes destructive interference at the receiver. This effect is known as *multipath fading* [19].

Unlike radio systems, multipath fading is not a major impairment in wireless optical transmission. The "antenna" in a wireless optical system is the light detector which typically has an active radiation collection area of approximately 1 cm². The relative size of this antenna with respect to the wavelength of the infrared light is immense, on the order of $10^4 \lambda$. The multipath propagation of light produces fades in the amplitude of the received electromagnetic signal at spacings on the order of half a wavelength apart. As mentioned earlier, the light detector is a square law device which integrates the square of the amplitude of the electromagnetic radiation impinging on it. The large size of the detector with respect to the wavelength of the light provides a degree of inherent spatial diversity in the receiver which mitigates the impact of multipath fading [2].

Although multipath fading is not a major impediment to wireless optical links, temporal dispersion of the received signal due to multipath propagation remains a problem. This dispersion is often modelled as a linear time invariant system since the channel properties change slowly over many symbol periods [1, 20]. The impact of multipath dispersion is most noticeable in diffuse infrared communication systems, which are described in more detail in Section 2.4.2. In short distance line-of-sight (LOS) links, presented in Section 2.4.1, multipath dispersion is seldom an issue. Indeed, channel models proposed for LOS links assume the LOS path dominates and model the channel as a linear attenuation and delay [21].

The modelling of the multipath response in a variety of indoor environments has been carried out to allow for computer simulation of communication

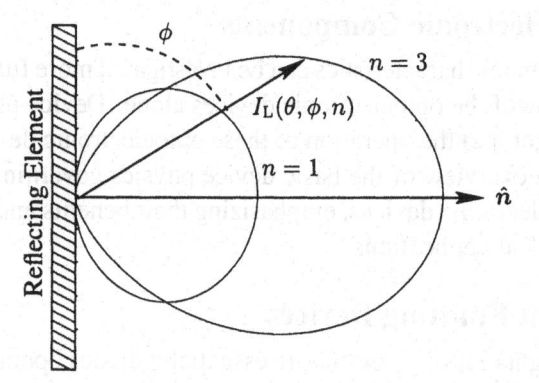

Figure 2.2. Example Lambertian radiation patterns for mode numbers $n = 1, 3$

systems. Gfeller and Bapst [1] introduced the concept of using diffuse optical radiation for indoor communication as well as defining the first simulation model. In their model, each surface in an indoor environment is partitioned into a set of reflecting elements which scatter incident optical radiation. A key assumption is that, regardless of the angle of incidence, each element scatters light with a *Lambertian* intensity pattern,

$$I_{\mathrm{L}}(\theta, \phi, n) = P_{\mathrm{total}}\frac{n+1}{2\pi}\cos^{n}\phi \qquad [\mathrm{W/sr}]$$

where P_{total} is the total reflected power, n is the *mode number* of the radiation pattern and angles $\phi \in [-\pi/2, \pi/2]$ and $\theta \in [0, 2\pi)$ are the polar and azimuthal angles respectively with respect to a normal, $\hat{n}$, to the reflecting element surface. The Lambertian optical intensity distribution is normalized so that integrating it over a hemisphere gives P_{total}. The mode number is a measure of the directivity of the reflected diffuse intensity distribution and typical values for plaster walls are near unity [1]. Figure 2.2 presents a plot of a cross-section of the Lambertian radiation pattern for $n = 1, 3$. Notice that this Lambertian radiation pattern models only diffuse reflections from surfaces and not specular reflections. In the Gfeller and Bapst model, the received power is simply the power from every element. There has been a continued interest in defining mathematical and simulation models for the multipath response of a variety of indoor settings. New, more accurate, analytic and simulation models have been developed which take into account multiple reflections as well as allow for fast execution time [22, 23, 21, 24, 25]. Additionally, experimental investigations have also been done to measure the response of a large number of channels and characterize the delay spread, path loss as well as investigating the impact of rotation [26, 20, 27]. Typical bandwidths for the multipath distortion is on the order of 10-50 MHz [11].

2.2   Optoelectronic Components

The basic channel characteristics can be investigated more fully by considering the operation of the optoelectronic devices alone. Device physics provides significant insight into the operation of these optoelectronic devices. This section presents an overview of the basic device physics governing the operation of certain optoelectronic devices, emphasizing their benefits and disadvantages for wireless optical applications.

2.2.1   Light Emitting Devices

Solid state light emitting devices are essentially diodes operating in forward bias which output an optical intensity approximately linearly related to the drive current. This output optical intensity is due to the fact that a large proportion of the injected minority carriers recombine giving up their energy as emitted photons.

To ensure a high probability of recombination events causing photon emission, light emitting devices are constructed of materials known as *direct band gap* semiconductors. In this type of crystal, the extrema of the conduction and valence bands coincide at the same value of wave vector. As a result, recombination events can take place across the band gap while conserving momentum, represented by the wave vector (as seen in Figure 2.3)[28]. A majority of photons emitted by this process have energy $E_{\text{photon}} = E_g = h\nu$, where E_g is the band gap energy, h is Planck's constant and ν is the photon frequency in hertz. This equation can be re-written in terms of the wavelength of the emitted photon as

$$\lambda = \frac{1240}{E_g} \qquad (2.3)$$

where λ is the wavelength of the photon in nm and E_g is the band gap of the material in electron-Volts. Commercial direct band gap materials are typically compound semiconductors of group III and group V elements. Examples of these types of crystals include: GaAs, InP, InGaAsP and AlGaAs (for Al content less than ≈ 0.45) [29].

Elemental semiconducting crystals silicon and germanium are *indirect band gap* materials. In these types of materials, the extrema of conduction and valence bands do not coincide at the same value of wave vector k, as shown in Figure 2.3. Recombination events cannot occur without a variation in the momentum of the interacting particles. The required change in momentum is supplied by collisions with the lattice. The lattice interaction is modelled as the transfer of phonon particles which represent the quantization of the crystalline lattice vibrations. Recombination is also possible due to lattice defects or due to impurities in the lattice which produce energy states within the band gap [29, 31]. Due to the need for a change in momentum for carriers to cross the

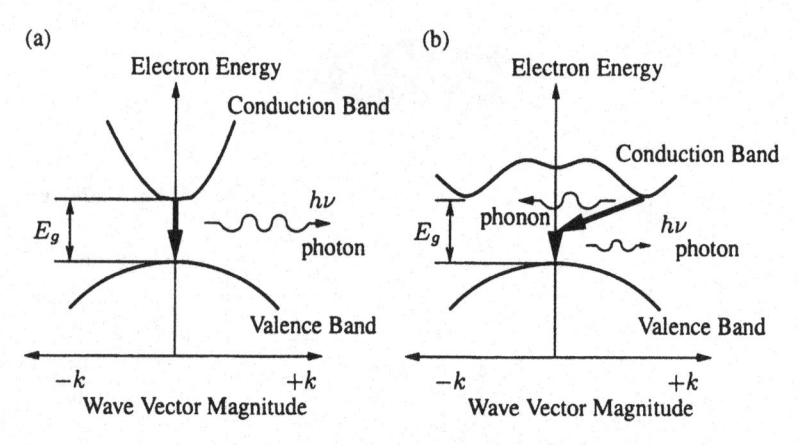

Figure 2.3. An example of a one dimensional variation of band edges with wave number (k) for (a) direct band gap material, (b) indirect band gap material (based on [30]).

band gap, recombination events in indirect band gap materials are less likely to occur. Furthermore, when recombination does take place, most of the energy of recombination process is lost to the lattice as heat and little is left for photon generation. As a result, indirect band gap materials produce highly inefficient light emitting devices [30].

The structure of light emitting devices fabricated in direct band gap III-V compounds greatly varies the properties of the emitted optical intensity signal. The two most popular solid-state light emitting devices are light emitting diodes (LEDs) and laser diodes (LDs).

Light Emitting Diodes

As was mentioned in Section 2.1.2, the use of the $780 - 950$ nm optical band is preferable due to the availability of low cost optoelectronic components. The direct band gap, compound semiconductor GaAs has a band gap of approximately 1.43 eV which corresponds to a wavelength of approximately 880 nm following (2.3).

Most modern LEDs in the band of interest are constructed as double heterostructure devices. This type of structure is formed by depositing two wide band gap materials on either side of a lower band gap material, and doping the materials appropriately to give diode action. A prototypical example of a double heterostructure LED is illustrated in Figure 2.4. Under forward bias conditions, the band diagram forms a potential well in the low band gap material (e.g., GaAs) into which carriers are injected. This region is known as the *active region* where recombination of the injected carriers takes place. The active region is flanked by properly doped higher band gap *confinement layers* (e.g., AlGaAs) which form a potential well confining the carriers. The recombination process in the active region occurs randomly and as a result the photons are

(a)

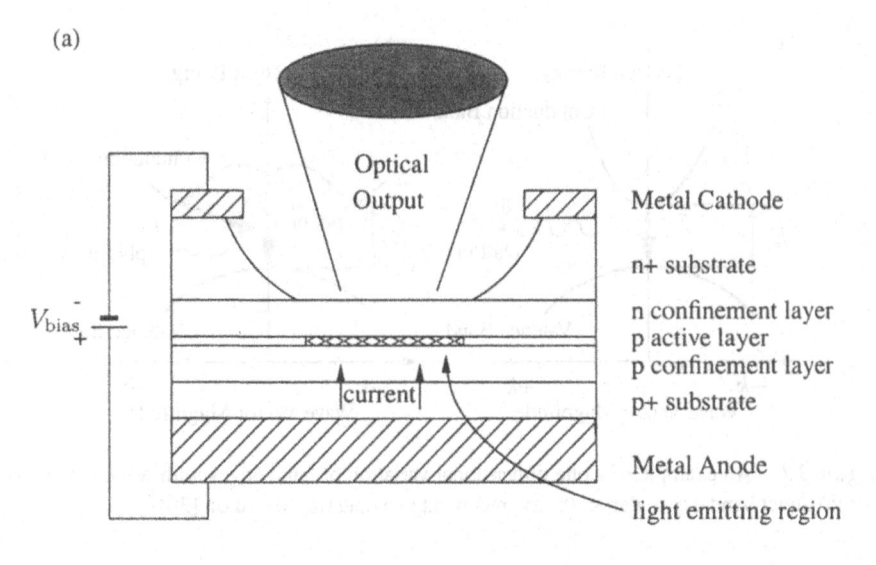

(b)

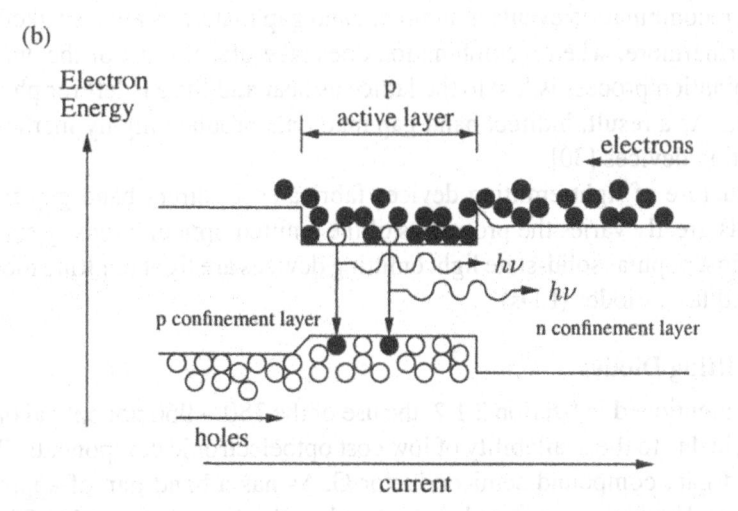

Figure 2.4.   An example of a double heterostructure LED (a) construction and (b) band diagram under forward bias (based on [29, 28]).

generated incoherently (i.e., the phase relationship between emitted photons is random in time). This type of radiation is termed *spontaneous emission* [29].

The advantages of using a double heterostructure stem from the fact that the injected carriers are confined to a well defined region. This confinement results in large concentration of injected carriers in the active region. This in turn reduces the radiative recombination time constant, improving the frequency response of the device. Another advantage of this carrier confinement is that the generated photons are also confined to a well defined area. Since the adjoining regions have a larger band gap than the active region, the losses due to absorption in these regions is minimized [28].

Using the structure for the LED in Figure 2.4, it is possible to derive an expression for the output optical power of the device as a function of the drive current as,

$$P_{\text{vol}} = h\nu \frac{J}{qd} B\tau_n \left(p_o + n_o + \frac{\tau_n J}{qd} \right), \tag{2.4}$$

where P_{vol} is the output power per unit device volume, J is the current density applied, $h\nu$ is the photonic energy, d is the thickness of the active region, B is the radiative recombination coefficient, τ_n is electron lifetime in the active region and p_o, n_o are carrier concentrations at thermal equilibrium in the active region [29].

Equation 2.4 shows that for low levels of injected current, $p_o \gg \tau_n J/qd$, P_{vol} is approximately proportional to the current density. As the applied current density increases (by increasing drive current) the optical output of the device exhibits more non-linear components. The choice of active region thickness, d, is a critical design parameter for source linearity. By increasing the thickness of the active region, the device has a wider range of input currents over which the behaviour is linear. However, an increase in the active region thickness reduces the confinement of carriers. This, in turn, limits the frequency response of the device as mentioned above. Thus, there is a trade-off between the linearity and frequency response of LEDs.

Another important characteristic of the LED is the performance of the device due to self-heating. As the drive current flows through the device, heat is generated due to the Ohmic resistance of the regions as well as the inefficiency of the device. This increase in temperature degrades the internal quantum efficiency of the device by reducing the confinement of carriers in the active region since a large majority have enough energy to surmount the barrier. This non-linear drop in the output intensity as a function of input current can be seen in Figure 2.5. The impact of self-heating on linearity can be improved by operating the device in pulsed operation and by the use of compensation circuitry [32–34]. Prolonged operation under high temperature environments reduces output optical intensity at a given current and can lead to device failure [35, 18].

The central wavelength of the output photons is approximately equal to the result given in (2.3). The typical width of the output spectrum is approximately 40 nm around the centre wavelength of 880 nm. This variation is due to the temperature effects as well as the energy distributions of holes and electrons in the active region [29].

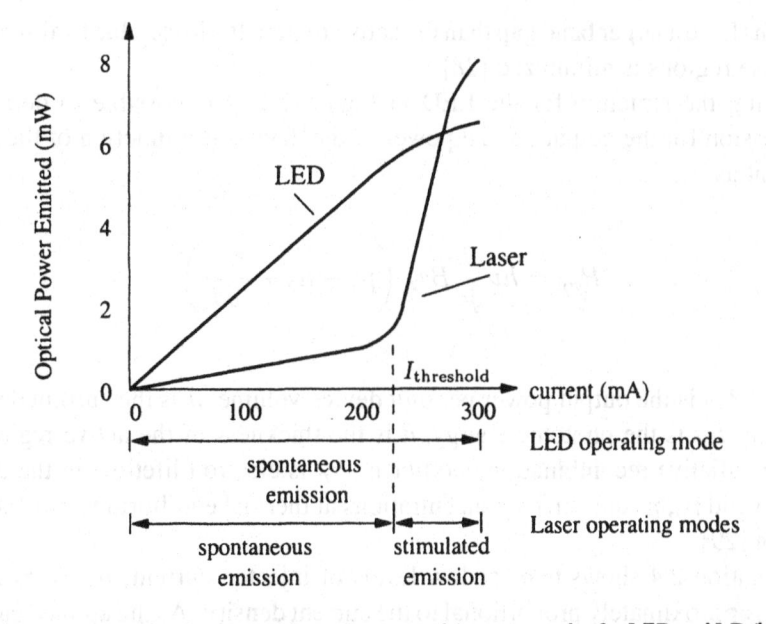

Figure 2.5. An example of an optical intensity versus drive current plot for LED and LD (based on [35])

Laser Diodes

Laser diodes (LDs) are a more recent technology which has grown from underlying LED fabrication techniques. LDs still depend on the transition of carriers over the band gap to produce radiant photons, however, modifications to the device structure allow such devices to efficiently produce coherent light over a narrow optical bandwidth.

As mentioned above, LEDs undergo spontaneous emission of photons when carriers traverse the band gap in a random manner. LDs exhibit a second form of photon generation process : *stimulated emission*. In this process, photons of energy E_g are incident on the active region of the device. In the active region, an excess of electrons is maintained such that in this region the probability of an electron being in the conduction band is greater than it being in the valence band. This state is called population inversion and is created by the confinement of carriers in the active region and the carrier pumping of the forward biased junction. The incident photon induces recombination processes to take place. The emitted photons in this process have the same energy, frequency, and phase as the incident photon. The output light from this reaction is said to be *coherent* [29, 30, 36].

In order for this process to be sustainable, the double heterostructure is modified to provide optical feedback. This optical feedback occurs essentially by placing a reflective surface to send generated photons back through the active region to re-initiate the recombination process. There are many techniques to provide this optical feedback, each with their merits and disadvantages. A Fabry-Perot laser achieves photon confinement by having internal reflection inside the active region. This is accomplished by adjusting the refractive-index of surrounding materials. The ends of the device have mirrored facets which are cleaved from the bulk material. One facet provides nearly total reflection while the other allows some transmission to free-space [29].

The operation of this optical feedback structure is analogous to microwave resonators which confine electromagnetic energy by high conductivity metal. These structures resonate at fixed set of modes depending on the physical construction of the cavity. As a result, due to the structure of the resonant cavity LDs emit their energy over a very narrow spectral width. Also, the resonant nature of the device allows for the emission of relatively high power levels.

Unlike LEDs which emit a light intensity approximately proportional to the drive current, lasers are threshold devices. As shown in Figure 2.5, at low drive currents, spontaneous emission dominates and the device behaves essentially as a low intensity LED. After the current surpasses the threshold level, $I_{\text{threshold}}$, stimulated emission dominates and the device exhibits a high optical efficiency as indicated by the large slope in the figure. In the stimulated emission region, the device exhibits an approximately linear variation of optical intensity versus drive current.

Comparison

The chief advantage of LDs over LEDs is in the speed of operation. Under conditions of stimulated emission, the recombination time constant is approximately one to two orders of magnitude shorter than during spontaneous recombination [28]. This allows LDs to operate at pulse rates in the gigahertz range, while LEDs are limited to megahertz range operation.

The variation of optical characteristics over temperature and age are more pronounced in LDs than in LEDs. As is the case with LEDs, the general trend is to have lower radiated power as temperature increases. However, a marked difference in LDs is that the threshold current as well as the slope of the characteristic can change drastically as a function of temperature or age of the device. For commercial applications of these devices, such as laser printers, copiers or optical drives, additional circuitry is required to stabilize operating characteristics over the life of the device [37, 38].

For LDs the linearity of the optical output power as a function of drive current above $I_{\text{threshold}}$ also degrades with device aging. Abrupt slope changes, known as *kinks*, are evident in the characteristic due to defects in the junction region as well as due to device degradation in time [35]. LEDs do not suffer from kinks over their lifetimes. Few manufactures quote linearity performance of their devices over their operating lifetimes.

Table 2.3. Comparison of LEDs versus LDs for wireless optical links.[*]

Characteristic	LED	LD
Optical Spectral Width	25–100 nm	0.1 to 5 nm
Modulation Bandwidth	Tens of kHz to Hundreds of MHz	Tens of kHz to Tens of GHz
Special Circuitry Required	None	Threshold and Temperature Compensation Circuitry
Eye Safety	Considered Eye Safe	Must be rendered eye safe
Reliability	High	Moderate
Cost	Low	Moderate to High

[*] based on [11, 18]

LDs are more difficult to construct and as a result can be more expensive than LEDs. As stated in Chapter 1, the use of inexpensive optical components is a key factor to ensuring the wide-spread adoption of wireless optical communications.

An important limitation for the use of LDs for wireless optical applications is the fact that it is necessary to render laser output eye safe. Due to the coherency and high intensity of the emitted radiation, the output light must be diffused. This requires the use of filters which reduce the efficiency of the device and increase system cost. LEDs are not optical point sources, as are LDs, and can launch greater radiated power while maintaining eye safety limits [11, 15]. Table 2.3 presents a comparison of the features of LDs and LEDs for wireless optical applications.

2.2.2   Photodetectors

Photodetectors are solid-state devices which perform the inverse operation of light emitting devices, i.e., they convert the incident radiant light into an electrical current. Photodetectors are essentially reverse biased diodes on which the radiant optical energy is incident, and are also referred to as *photodiodes*. The incident photons, if they have sufficient energy, generate free electron-hole pairs. The drift or diffusion of these carriers to the contacts of the device constitutes the detected photocurrent.

Inexpensive photodetectors can be constructed of silicon (Si) for the 780–950 nm optical band. The photonic energy at the 880 nm emission peak of GaAs is approximately $E_g = 1.43$ eV, by rearranging (2.3). Since the band gap of silicon is approximately 1.15 eV, these photons have enough energy to promote electrons to the conduction band, and hence are able to create free electron-hole pairs. Figure 2.6 shows that the sensitivity of a silicon photodiode is maximum in the optical band of interest.

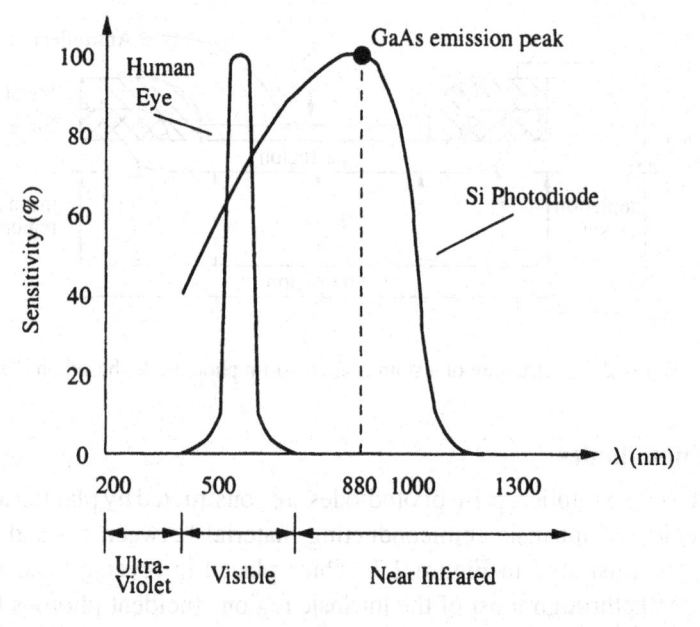

Figure 2.6. Relative sensitivity curve for a silicon photodiode (based on [31]). Note that the position of the GaAs emission line is located near the peak in sensitivity of the photodiode.

The basic steady-state operation of a solid-state photodiode can be modelled by the expression,

$$I_p = q\eta_i \frac{P_p}{h\nu}, \tag{2.5}$$

where I_p is the average photocurrent generated, η_i is the internal quantum efficiency of the device, P_p is the incident optical power and $h\nu$ is the photonic energy. The internal quantum efficiency of the device, η_i, is the probability of an incident photon generating an electron-hole pair. Typical values of η_i range from 0.7 to 0.9. This value is less than 1 due to current leakage in the device, absorption of light in adjacent regions and device defects [18].

Equation (2.5) can be re-arranged to yield the *responsivity* of the photodiode in the following manner,

$$R_p = \frac{I_p}{P_p} = \frac{q\eta_i}{h\nu} \tag{2.6}$$

The units of responsivity (R_p) are in amperes per watt, and it represents the optoelectronic conversion factor from optical to electrical domain. Responsivity is a key parameter in photodiode models, and is taken at the central optical frequency of operation.

Two popular examples of photodiodes currently in use include p-i-n photodiodes and avalanche photodiodes.

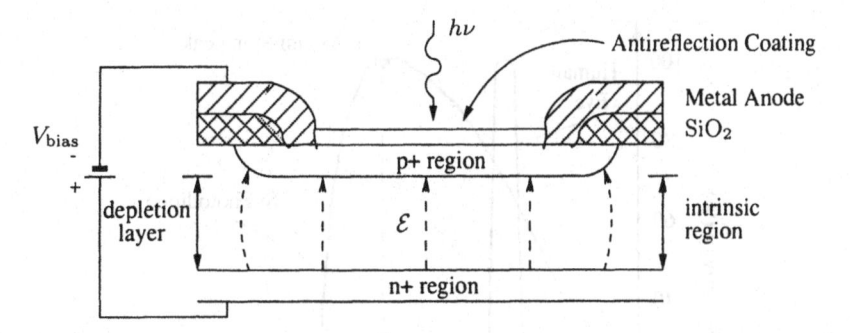

Figure 2.7.   Structure of a simple silicon p-i-n photodiode (based on [39]).

p-i-n Photodiodes

As the name implies, p-i-n photodiodes are constructed by placing a relatively large region of intrinsic semiconducting material between p+ and n+ doped regions as illustrated in Figure 2.7. Once placed in reverse bias, an electric field extends through most of the intrinsic region. Incident photons first arrive upon an anti-reflective coating which improves the coupling of energy from the environment into the device. The photons then proceed into the p+ layer of the diode. The thickness of the p+ layer is made much thinner than the absorption depth of the material so that a majority of the incident photons arrive in the intrinsic region. The incident light is absorbed in the intrinsic region, producing free carriers. Due to the high electric field in this region ($\mathcal{E}$), these carriers are swept up, and collected across the junction at a saturation velocity on the order of 10^7 cm/s. This generation and transport of carriers through the device is the origin of the photocurrent.

Although carrier transit time is an important factor limiting the frequency response of photodiodes for fibre applications, the main limiting factor for wireless applications is the junction capacitance of the device. In wireless applications, devices must be made with relatively large areas so as to be able to collect as much radiant optical power as possible. As a result, the capacitance of the device can be relatively large. Additionally, the junction capacitance is increased due to the fact that in portable devices with battery power supplies low reverse bias voltages are available. Typical values for this junction depletion capacitance at a reverse bias of 3.3 V range from 2 pF for expensive devices used in some fibre applications to 20 pF for very low speed, and cost devices. Careful design of receiver structures is necessary so as not to unduly reduce system bandwidth or increase noise [40].

The relationship between generated photocurrent and incident optical power for p-i-n photodiodes in (2.5) has been shown to be linear over six to eight decades of input level [41, 31]. Second order effects appear when the device is operated at high frequencies as a result of variations in transport of carriers through the high-field region. These effects become prevalent at frequencies above approximately 5 GHz and do not limit the linearity of links at lower frequencies of operation [42]. Since the frequency of operation is limited due to junction capacitance, the non-linearities due to charge transport in the device are typically not significant. The p-i-n photodiode behaves in an approximate linear fashion over a wide range of input optical intensities.

Avalanche Photodiodes

The basic construction of avalanche photodiodes (APDs) is very similar to that of a p-i-n photodiode. The difference is that for every photon which is absorbed by the intrinsic layer, more than one electron-hole pair may be generated. As a result, APDs have a photocurrent gain of greater than unity, while p-i-n photodiodes are fixed at unit gain.

The process by which this gain arrives is known as *avalanche multiplication* of the generated carriers. A high intensity electric field is established in the depletion region. This field accelerates the generated carriers so that collisions with the lattice generate more carriers. The newly generated carriers are also accelerated by the field, repeating the impact generation of carriers. The photocurrent gain possible with this type of arrangement is of the order 10^2 to 10^4 [41, 39]. In wired fibre networks, the amplifying effect of APDs improves the sensitivity of the receiver allowing for longer distances between repeaters in the transmission network [28].

The disadvantage of this scheme is that the avalanche process generates excess shot noise due to the current flowing in the device. This excess noise can degrade the operation of some free space links since a majority of the noise present in the system is due to high intensity ambient light. These noise sources are discussed in more detail in Section 2.3.

The avalanche gain is a strong non-linear function of bias voltage and temperature. The primary use of these devices is in digital systems due to their poor linearity. Additional circuitry is required to stabilize the operation of these devices. As a result of the overhead required to use these devices, the system reliability may also be degraded [18].

Comparison

APDs provide a gain in the generated photocurrent while p-i-n diodes generate at most one electron-hole pair per photon. It is not clear that this gain produces an improvement in the signal-to-noise ratio (SNR) in every case. Indeed, for the case of a free space optical link operating in ambient light, APDs can actually provide a decrease in SNR [2], as described in Section 2.3.

Due to the non-linear dependence of avalanche gain on the supply voltage and temperature, APDs exhibit non-linear behaviour throughout their operating regime. The addition of extra circuitry to improve this situation increases

Table 2.4. Comparison of p-i-n photodiodes versus avalanche photodiodes for wireless optical links.*

Characteristic	p-i-n Photodiode	Avalanche Photodiode
Modulation Bandwidth (ignoring circuit)	Tens of MHz to Tens of GHz	Hundreds of MHz to Tens of GHz
Photocurrent Gain	1	$10^2 - 10^4$
Special Circuitry Required	None	High Bias Voltages and Temperature Compensation Circuitry
Linearity	High	Low – suited to digital applications
Cost	Low	Moderate to High

* Based on [18, 41].

cost and lowers system reliability. Additional circuitry is also necessary to generate the high bias voltages necessary for high field APDs. Typical supply voltages range from 30 V for InGaAs APDs to 300 V for silicon APDs. Since these devices are destined for portable devices with limited supplies, APDs are not appropriate for this application. Table 2.4 presents a summary of the characteristics of p-i-n photodiodes and APDs.

There is a large number of available p-i-n diodes at relatively low cost and at a variety of wavelengths. They have nearly linear optoelectronic characteristics over many decades of input level. Unlike APDs, p-i-n photodiodes can be biased from lower supplies with the penalty of increasing junction capacitance. Figure 2.8 in Section 2.3 illustrates the typical circuit model used for representing the impact of front-end photodiode capacitance. Table 2.5 presents the responsivities and gain of p-i-n and APD devices manufactured in a variety of materials. In contrast to APD structures, the p-i-n diodes have smaller values of responsivity and a photocarrier multiplier gain of unity.

As mentioned earlier, most commercial indoor wireless optical links employ inexpensive Si photodetectors and LEDs in the 850-950 nm range. However, some long-range, outdoor free-space optical links employ compound photodiodes operating at longer wavelength to increase the amount of optical power transmitted while satisfying eye-safety limits. Additionally, these long-range links also employ APD receivers to increase the sensitivity of the receiver [13]. Care must be taken in the selection of photodiode receivers to ensure that cost, performance and safety requirements are satisfied.

Table 2.5. Characteristics of p-i-n and APD photodiodes fabricated from a variety of materials.*

Material & Structure	Wavelength (nm)	Responsivity (A/W)	Gain
Si p-i-n	300–1100	0.5	1
Ge p-i-n	500–1800	0.7	1
InGaAs p-i-n	1000–1700	0.9	1
Si APD	400–1000	77	150
Ge APD	800–1300	7	10
InGaAs APD	1000–1700	9	10

* From [13].

2.3 Noise

Along with specifications regarding the frequency and distortion performance, the noise sources of a wireless optical link are critical factors in determining performance. As is the case in nearly all communication links, the determination of noise sources at the input of the receiver is critical since this is the location where the incoming signal contains the least power.

As discussed in Section 2.2.2, p-i-n photodiodes are commonly used as photodetectors for indoor wireless infrared links. The two primary sources of noise at the receiver front end are due to noise from the receive electronics and shot noise from the received DC photocurrent.

As is the case with all electronics, noise is generated due to the random motion of carriers in resistive and active devices. A major source of noise is thermal noise due to resistive elements in the pre-amplifier. If a low resistance is used in the front end to improve the frequency response, an excessive amount of thermal noise is added to the photocurrent signal. Transimpedance pre-amplifiers provide a low impedance front end through negative feedback and represent a compromise between these constraints [40]. Figure 2.8 illustrates a schematic of a front end with photodetector as well as noise sources indicated. Thermal noise is generated independently of the received signal and can be modelled as having a Gaussian distribution. This noise is shaped by a transfer function dependent on the topology of the pre-amplifier once the noise power is referred to the input of the amplifier. As a result, circuit noise is modelled as being Gaussian distributed and, in general, non-white [11].

Photo-generated shot noise is a major noise source in the wireless optical link. This noise arises fundamentally due to the discrete nature of energy and charge in the photodiode. Carrier pairs are generated randomly in the space charge region due to the incident photons. Furthermore, carriers traverse the potential barrier of the p-n junction in a random fashion dependent on their energy. The

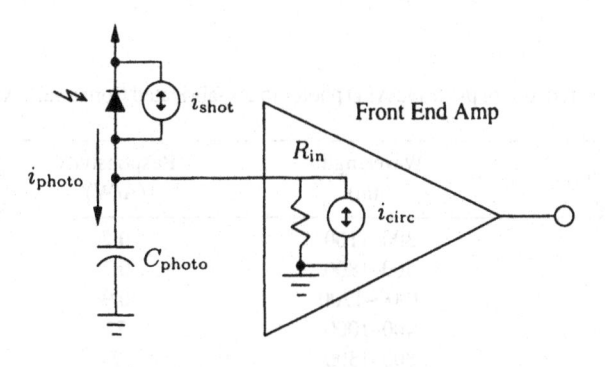

Figure 2.8.   Diagram of a front-end photodiode detector along with channel impairments.

probabilistic generation and transport of carriers due to quantum effects in the photodiode gives rise to shot noise in the photocurrent, as illustrated in Figure 2.8. This random process can be modelled as having Poisson distribution with a white power spectral density [40, 43].

Using these two sources of noise, the signal-to-noise ratio for a wireless optical link can be approximated for a simple example (based on [44]). The input signal to the receiver is a time varying optical intensity signal. Let the transmitted intensity signal, $x(t)$, be a fixed sinusoid of the form

$$x(t) = P_t(1 + m \sin \omega t),$$

where P_t is the average transmitted power and m is the amplitude of the sinusoid. To ensure transmission is possible, $|m| \leq 1$, since negative intensity values are not possible as discussed in Section 2.1.

Assume that the intensity signal is only attenuated as it propagates through free-space to the receive side. The received signal also consists of ambient, background light that is present in the channel and detected by the photodiode. This ambient light consists of incandescent light, natural light and other illumination in the environment. The received intensity, $r(t)$, can be written as

$$r(t) = P_r(1 + m \sin \omega t) + P_{\text{amb}},$$

where P_r is the average power at the receive side and P_{amb} is the power of the ambient light incident on the photodiode. The photodiode converts this incident optical intensity into a photocurrent in accordance with the responsivity relationship in (2.6). The signal and DC quantities of the photocurrent can be isolated in the following form :

$$\begin{aligned} i_{\text{photo}}(t) &= R_p \cdot r(t) \\ &= R_p(P_r + P_{\text{amb}}) + (R_p P_r m \sin \omega t). \end{aligned}$$

The electrical signal power at the receive side is contained entirely in the time varying component and can be written as

$$P_{signal} = \frac{1}{2}m^2(R_pP_r)^2.$$

The photo-generated shot noise at the receiver arises due to both the ambient light and the transmitted signal. Since $P_{amb} \gg P_r$, consider only the DC component of the received photocurrent. Since the noise power due to the pre-amplifier, $\overline{i^2_{circ}}$, and due to shot noise, $\overline{i^2_{shot}}$, is uncorrelated, the total noise power is simply,

$$\begin{aligned} P_{noise} &= \overline{i^2_{circ}} + \overline{i^2_{shot}} \\ &= \overline{i^2_{circ}} + 2qR_p(P_r + P_{amb})B_{eff}, \end{aligned}$$

where q is the electronic charge and B_{eff} is the equivalent noise bandwidth of the system. Combining the results, an estimate of the signal-to-noise ratio of the system can be formed as,

$$\begin{aligned} SNR_{link} &= \frac{P_{signal}}{P_{noise}} \\ &= \frac{1}{2} \cdot \frac{m^2(R_pP_r)^2}{\overline{i^2_{circ}} + 2qR_p(P_r + P_{amb})B_{eff}}. \end{aligned} \qquad (2.7)$$

The dominant source of noise in a wireless optical channel is due to the ambient background light. To reduce the impact of ambient light, optical filters can be used to attenuate lower wavelength visible and higher frequency light sources with little added cost [1]. In some links, this ambient light may be as much as 25 dB greater than the signal power, even after optical filtering [2]. Many wireless optical links operate in this shot-noise limited regime due to the intense background illumination. In these cases, the ambient light shot noise component dominates the shot noise due to the received signal as well as the circuit noise. Thus, for a shot-noise limited links, (2.7) to be simplified to,

$$SNR_{link} \approx \frac{R_pm^2P_r^2}{4qP_{amb}B_{eff}}.$$

Using this assumption, the resulting noise of the channel is signal independent, white shot noise following a Poisson distribution. This high intensity shot noise is the result of the summation of many independent, Poisson distributed random variables. In the limit, as the number of random variables summed approaches infinity, the cumulative distribution function approaches a Gaussian distribution by the central limit theorem. Thus, the dominant noise source in many indoor wireless optical channels can be modelled as being white, signal independent and having a Gaussian distribution [43, 11]. In a more mathematically rigorous fashion, it is possible to show that the moment generating function of high intensity shot noise approaches a Gaussian distribution at high intensities [45]. Detailed physical studies [46, 47] of avalanche photodiodes indicate that the noise probability density, $f_n(x)$, can be modelled as

$$f_n(x) = \frac{1}{\sqrt{2\pi}\sigma_n(1 + x/\lambda)^{\frac{3}{2}}} \exp\left(-\frac{x^2}{2\sigma_n^2(1 + x/\lambda)}\right),$$

where σ_n^2 is a device dependent parameter proportional to the average optical power and λ is proportional to the received optical intensity and inversely dependent on excess noise factor of the avalanche photodiode. In the case of photodiodes under high illumination and small excess noise factors, which is the case for silicon photodiodes, λ is large. In this case, the resulting noise distribution tends to a Gaussian distribution [46, 48].

The characteristics of the noise depend on the configuration of the link. The *field-of-view* (FOV) of a receiver is the extent of solid-angles from which light is detected at the receiver [3]. The received high intensity optical radiation impinging on the photodiode is both from ambient lighting sources and from the transmitted optical intensity signal. Narrow FOV links are able to reject a large component of ambient light. The resulting noise can still be modelled as being Gaussian distributed but dependent on the transmitted signal. Chapter 7 describes one such channel, the pixelated wireless optical channel in which the noise variance is modelled as varying linearly with the transmitted optical signal. In the case of wide FOV receivers, the ambient light dominates the received signal. As a result, the noise at the receiver can be approximated as being independent of the transmitted data. In this case, the noise is modelled as additive, signal independent, white, Gaussian with zero mean and variance σ_n^2 [11].

This situation is in contrast to optical fibres where the ambient light is essentially zero, and circuit noise is the dominant noise factor [44, 45]. The use of an APD is advantageous in fibre applications as long as the circuit noise is much greater than the added shot noise of the APD. In this manner, APDs can provide a gain to the signal portion of the received power while keeping the noise power essentially constant. The net effect is to allow for wider repeater spacing in a fibre network, reducing system cost [28].

Emissions from fluorescent lighting create a noise source unique to wireless optical channels. Fluorescent lamps have strong emissions at the spectral lines of argon in the 780–950 nm near infrared band. Although economical narrow band optical filters have been used for some time in such links [1], significant energies are still detected by the photodiode. The detected output of fluorescent lamps is nearly deterministic and periodic with components at multiples of the ballast drive frequency. Most fluorescent ballasts drive the lamps at the line

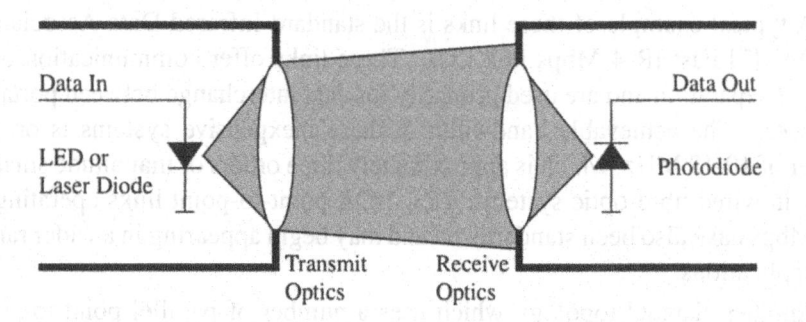

Figure 2.9. A point-to-point wireless optical communications system.

frequency of 50–60 Hz, with harmonics up to tens of kilohertz. Modern ballasts modulate the lamp at higher frequencies to improve power efficiency and reduce unit size. Typical modulation rates are 22 kHz and 45 kHz. The harmonics generated by these sources, and detected by the photodiode extend into the hundreds of kilohertz and can present an impediment to wireless optical data transmission. The impact of periodic interference from high frequency modulated fluorescent light sources has only recently been investigated for wireless optical links [49, 11, 2]. In this work it is assumed that the additive white noise is dominant over fluorescent light interference. However, the robustness of certain key modulation schemes against fluorescent light sources is briefly discussed in Chapter 3.

2.4 Channel Topologies

The characteristics of the wireless optical channel can vary significantly depending on the topology of the link considered. This section presents three popular wireless optical channel topologies and discusses the channel characteristics of each.

2.4.1 Point-to-Point Links

Point-to-point wireless optical links operate when there is a direct, unobstructed path between a transmitter and a receiver. Figure 2.9 presents a diagram of a typical point-to-point wireless optical link. A link is established when the transmitter is oriented toward the receiver. In narrow field-of-view applications, this oriented configuration allows the receiver to reject ambient light and achieve high data rates and low path loss. The main disadvantage of this link topology is that it requires pointing and is sensitive to blocking and shadowing. The frequency response of these links is limited primarily by front-end photodiode capacitance. Since inexpensive, large-area photodiodes are typically used with limited reverse bias, the depletion capacitance significantly limits the link bandwidth [50, 51].

A typical example of these links is the standard Infrared Data Association (IrDA) [7] Fast IR 4 Mbps link [52]. These links offer communication over 1 m of separation and are used primarily for data interchange between portable devices. The achievable bandwidth in these inexpensive systems is on the order of 10-12 MHz, which is approximately three orders of magnitude smaller than in wired fibre-optic systems. New IrDA point-to-point links operating at 16 Mbps have also been standardized and may begin appearing in a wider range of applications.

Another channel topology which uses a number of parallel point-to-point links is the *space division multiplexing* architecture. Space division multiplexing is a technique by which a transmitter outputs different data in different spatial directions to allow for the simultaneous use of one wavelength by multiple users. In one such system, a ceiling-mounted basestation has a number of narrow beams establishing point-to-point links in a variety of directions in a room. A fixed receiver, once aligned to within 1° of a transmitter beam, establishes a high speed link at up to 50 Mb/s [53]. Another means of implementing a space division multiplexing system is to use a tracked optical wireless architecture. In this system, the transmitter beams are steerable under the control of a tracking subsystem. Tracking is typically accomplished by a "beacon" LED or FM transmitter on the mobile terminal. These systems are proposed to provide 155 Mb/s ATM access to mobile terminals in a room [54–56]. Electronic tracking systems have also been proposed which exploit a diffuse optical channel to aid in acquisition [57]. The advantage of this topology is that it is extremely power efficient and supports a large aggregate bandwidth inside of a room at the expense of system complexity.

Point-to-point wireless optical links have been implemented in a wide variety of short- and long-range applications. Short range infrared band links are being designed to allow for the transfer of financial data between a PDA or cellphone and a point-of-sale terminal [58, 59]. Wireless optical links are chosen as the transmission medium due to the low cost of the transceivers and the security available by confining optical radiation. The IrDA has specified a standard for this financial application under the title IrDAFM (financial messaging) [7]. Medium range indoor links have also been developed to extend the range of Ethernet networks in an office environment. A 10 Mbps point-to-point wireless infrared link to extend Ethernet networks has been deployed over a range of at most 10 m [60]. Higher rate, 100 Mbps point-to-point wireless infrared links have also been designed to extend Ethernet networks in indoor environments [61].

A related wireless optical link, not treated explicitly in this book, is the long range outdoor optical link. Using more expensive transmitters and receivers and pointing mechanisms, multi-gigabit per second transmission is possible over 4 km [62, 14]. Ultra-long range point-to-point links are also being con-

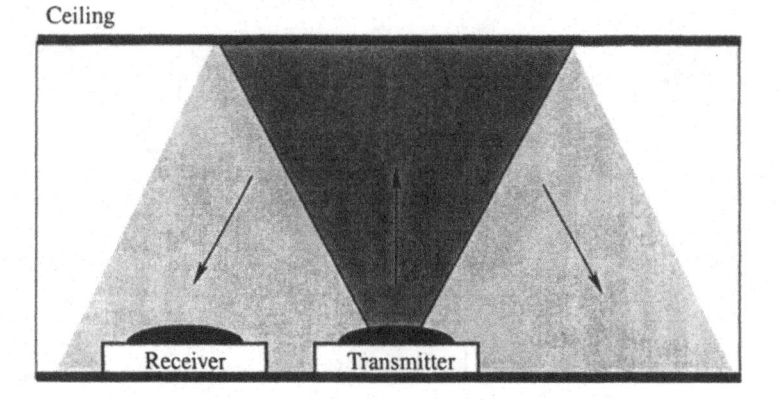

Figure 2.10. A diffuse wireless optical communications system.

sidered for earth-to-space communications at rates in excess of 1 Mbps [63]. Indeed, the most fantastic long-range, point-to-point wireless optical link may be the one investigated by the optical SETI project which extends the search for extraterrestrial intelligence to include wireless optical signals [64] !

2.4.2   Diffuse Links

Diffuse transmitters radiate optical power over a wide solid angle in order to ease the pointing and shadowing problems of point-to-point links. Figure 2.10 presents a block diagram of a diffuse wireless optical system. The transmitter does not need to be aimed at the receiver since the radiant optical power is assumed to reflect from the surfaces of the room. This affords user terminals a wide degree of mobility at the expense of a high path loss. These channels, however, suffer not only from optoelectronic bandwidth constraints but also from low-pass multipath distortion [1, 2, 11]. Unlike radio frequency wireless channels, diffuse channels do not exhibit fading. This is due to the fact that the receive photodiode integrates the optical intensity field over an area of millions of square wavelengths, and hence no change in the channel response is noted if the photodiode is moved a distance on the order of a wavelength [20, 11]. Thus, the large size of the photodiode relative to the wavelength of light provides a degree of spatial diversity which eliminates multipath fading.

Multipath distortion gives rise to a channel bandwidth limit of approximately 10-200 MHz depending on room layout, shadowing and link configuration [11, 21, 20, 27]. Many channel models based on measurements allow for the accurate simulation of the low-pass frequency response of the channel [27, 24, 22, 21, 20].

The IrDA and the IEEE have similar standards for diffuse infrared links. The IrDA Advanced Infrared (AIr) standard allows communication at rates up to

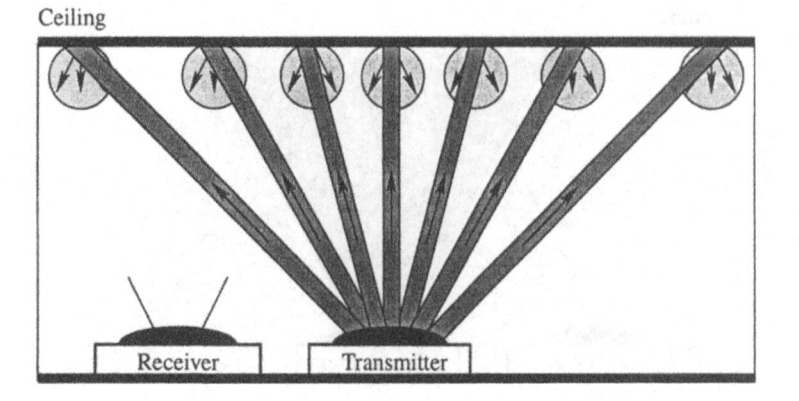

Figure 2.11. A quasi-diffuse wireless optical communications system.

4 Mbps with repetition coding [65, 7]. The IEEE wireless infrared standard falls under the 802.11 standard and allows diffuse transmission at a maximum of 2 Mbps [66, 6]. Both systems used pulse-position modulation (PPM) which is a coded version of on-off keying. Experimental indoor wireless optical links have been demonstrated at 50 Mbps using on-off keying over a horizontal range of approximately 3 m [23]. A commercial indoor diffuse wireless optical link aimed at digital audio and set-top box applications claims data rates of up to 5 Mbps in typical indoor environments [67]. An early diffuse wireless optical system was employed in a portable computer called PARCTAB, developed at the Xerox Palo Alto Research Center (PARC) in 1993 [68]. The diffuse link was able to provide data rates of up to 19.2 kbps and was used to communicate electronic mail and other data to a hand-held computing device.

Diffuse wireless links have also been employed in so called *smart badges* which are used to provide location information for individuals in a building or complex. The most notable of such systems is the Active Badge system developed by Olivetti Research Labs from 1989-1992 [69, 70]. In this system, individuals wear identification cards equipped with infrared transmitters which transmit a unique code for each individual. Receivers in each room detect the emissions and transmit the information to a host. The location information is used to automatically route calls for the user to the nearest telephone, to allow terminal access and identification purposes.

2.4.3   Quasi-Diffuse Links

Quasi-diffuse links inherit aspects of both point-to-point and diffuse links to optimize link throughput [71, 72]. Figure 2.11 presents a diagram of a prototypical quasi-diffuse link.

The transmitter illuminates the ceiling with a series of slowly diverging beam sources which illuminate a grid of spots on the ceiling. In experimental settings, these multiple beams are created using individual light sources [71, 73] and proposed techniques using holographic beam splitters appear promising [74]. The transmit beams suffer a small path loss nearly independent of the length of the link from the transmitter to the ceiling due to the low beam divergence [71]. The data transmitted on all beams is identical. The receiver consists of multiple concentrator/photodiode pairs, each with a non-overlapping, narrow FOV of the ceiling. The FOV of each receiver is typically set to see at least one spot on the ceiling. These narrow FOV receivers reject a majority of multipath distortion and provide a link with an improved bandwidth although the link is more sensitive to shadowing relative to diffuse links. Spatially localized interferers, such as room illumination, can be rejected by using the spatial diversity of the multiple receivers. In a diffuse scheme all the noise power is collected along with the signal power.

2.4.4 Comparison

Table 2.6 presents a comparison of some of the characteristics of the three channel topologies discussed. The point-to-point topology is a low complexity means to achieve high data rate links at the expense of mobility and pointing requirements. Diffuse links suffer from high path loss but offer a great degree of mobility and robustness to blocking. Quasi-diffuse links permit higher data rates by requiring users to aim their receivers at the ceiling but suffer from a higher implementation cost due to the multi-beam transmitter. Thus, each channel topology is suited to a different application depending on required data rates and channel conditions.

It may also be advantageous to combine the operation of the various topologies to form a more robust link. Recent work has demonstrated experimental configurations which use a diffuse wireless optical channel to aid in acquiring tracking and to serve as a backup link to improve user mobility [57].

2.5 Summary

This chapter has introduced the primary physical constraints which are imposed on any signalling scheme for the optical intensity channel. The amplitude non-negativity constraint follows directly from the fact that optical intensity modulators are employed. Eye and skin safety considerations lead to an average optical power, or equivalently amplitude, constraint. These constraints are significantly different than the electrical power constraints found in conventional channels. Indoor wireless optical channels also suffer from multipath distortion due to the response of a room. Typical optoelectronics used in these links are LEDs and silicon p-i-n photodiodes in the 800-900 nm wavelength

Table 2.6. Comparison of wireless optical topologies.

	Point-to-Point	Diffuse	Quasi-Diffuse
Rate	High	Low-Moderate	Moderate
Pointing Required	Yes	No	Somewhat
Immunity to Blocking	Low	High	Moderate-High
Mobility	Low	High	Moderate-High
Complexity of Optics	Low	Low-Moderate	High
Ambient Light Rejection	High	Low	High
Multipath Distortion	None	High	Low
Path Loss	Low	High	Moderate

Table 2.7. Some typical values for an experimental short-range wireless optical link.*

Measurement	Value	Unit
3 dB Bandwidth	35	MHz
Second Order Harmonic Distortion	-39	dBc
SFDR	23	dB

* Based on [5].

band. These devices provide a good trade-off between cost, safety and ease of implementation. Point-to-point, diffuse and quasi-diffuse channel topologies provide a range of channels with varying degrees of mobility and data rate.

Table 2.7 presents some results of a short-range wireless optical channel used to characterize some commercial optoelectronic components. A Mitel 1A301 LED [75] and a Temic BPV10NF [76] silicon p-i-n photodiode were chosen as test subjects. The Mitel LED is designed for 266 Mbps fibre links, and has a reported bandwidth of 350 MHz. The Temic photodiode is reported as having a bandwidth of 100 MHz, however, the measurement method is not well documented. The suggested applications for this photodiode are for 450 kHz/1.3 MHz FSK remote control purposes as well as 4 Mbps IrDA links.

The distance between transmitter and receiver was set to 1.5 cm and electronics were designed so that the characteristics of the devices could be measured. The results indicate that the channel has a 3 dB bandwidth of 35 MHz, limited by the photodiode. The response exhibits a single pole drop-off up to approximately 100 MHz after which point the drop-off falls more abruptly. The second harmonic falls 39 dB below the fundamental and is limited by the experimental setup. The *spurious free dynamic range* (SFDR) is a measure of the range of output power over which the channel can be used as a linear one. At low

Table 2.8. Typical values for a measured indoor diffuse wireless optical link.*

Measurement	Value	Unit
Path Loss (unshadowed)	53–63	dB (optical)
Path Loss (shadowed)	55–67	dB (optical)
Multipath Induced Bandwidth	20–40	MHz

* Based on [20].

output levels, the noise floor limits performance, while at high output levels distortion products exceed the noise floor and become significant at the output. The measured SFDR, over the bandwidth of the channel of the channel is 23 dB [5]. These measurements suggest that over a wide range of input powers the optoelectronics behave nearly linearly and allow for analog signal transmission. Additionally, the use of large photodiodes and inexpensive LEDs limits the bandwidth and care must be taken in signalling design to efficiently exploit the channel.

Table 2.8 presents some measured values from an experimental diffuse channel [20]. The area of the detector is 1 cm^2 and the transmitter and receiver are both directed upwards in a variety of office settings. The key feature to note is that diffuse wireless optical channels are *bandwidth-limited* due to the multipath response of the room and also suffer from a large path loss, especially when shadowed [20].

In subsequent chapters, the challenge of signalling design for bandwidth constrained wireless optical channels will be considered. Various channel models will be used to represent the amplitude, bandwidth and optical power constraints inherent to both point-to-point and diffuse channels.

Chapter 3

AN INTRODUCTION TO
OPTICAL INTENSITY SIGNALLING

The wireless optical intensity channel requires that all transmitted signals assume non-negative values and that the average amplitude, i.e., optical power, is bounded. Any signal which is transmitted is also corrupted by a linear low pass frequency response due to both optoelectronics and multipath distortion. Random shot noise, due to ambient lighting, is the dominant source of noise and must be considered in any signalling design for the channel.

It is often not possible, or highly inefficient, to apply signalling techniques designed for radio frequency channels directly to optical intensity channels since they do not take into account the constraints of the channel. The purpose of this chapter is to present the issues involved in the design of signalling for wireless optical channels. A mathematical channel model is given which represents the main features of an indoor wireless optical channel to allow for modem design. In later chapters, the model will be refined to a wider class of channels. Basic communication design concepts are briefly reviewed and linked to optical intensity channels. A brief overview of popular signalling techniques is included along with a comparison of binary level and multilevel modulation. As is the case in conventional electrical channels, multi-level modulation is shown to provide a gain in spectral efficiency at the cost of power efficiency.

The goal of this chapter is to establish the framework for signalling design and to pose the fundamental communications design problem for the wireless optical intensity channel.

3.1   Communication System Model

In order to proceed with system design, a mathematical model of the channel is required. The mathematical model should be based on the physical principle of the channel and should demonstrate the dominant features of the channel while remaining simple enough to permit analysis. The channel model will vary depending on channel conditions as well as on the channel topology. The channel model presented in this section is a simple model for a fixed point-to-point or diffuse channel. In Chapter 7, a more general model will be introduced in the case of multiple transmitters and receivers with narrow field-of-views.

3.1.1 Channel Model

As discussed in Chapter 2, the opto-electrical conversion is typically performed by a silicon photodiode which performs direct-detection of the incident optical intensity signal. In the far-field case, the channel response from transmitted intensity, $I(t)$, to the receive photocurrent, $y(t)$, in Fig. 2.1 is well approximated as

$$y(t) = r\frac{I(t)}{D^2} \otimes h(t) + n(t),$$

where $\otimes$ denotes convolution, r is the detector sensitivity in units of $A \cdot m^2/W$, D is the distance between transmitter and receiver, $n(t)$ is the noise process and $h(t)$ is the channel response [20, 21, 11–1]. Without loss of generality, let the $1/D^2$ factor be lumped with $h(t)$ to yield

$$y(t) = rI(t) \otimes h(t) + n(t). \tag{3.1}$$

The additive noise, $n(t)$, is modelled as additive, signal independent, white, Gaussian with zero mean and variance σ^2. This model for the noise is realistic for wide field-of-view point-to-point links and for diffuse links in the presence of intense ambient illumination [2, 11].

As discussed in Section 2.2.1, LEDs and laser diodes above threshold perform a near linear conversion between the input drive current and the output optical intensity. The electro-optical conversion can be modelled as $I(t) = gx(t)$, where g is the optical gain of the device in units of $W/(A \cdot sr)$. Substituting $I(t) = gx(t)$ into (3.1) gives

$$y(t) = rg \cdot x(t) \otimes h(t) + n(t).$$

Without loss of generality, we may set $rg = 1$ to simplify analysis. In this manner, the free space optical channel is represented by a baseband electrical model.

As discussed in Section 2.1, the channel response $h(t)$ is typically low-pass due to photodiode capacitance and multipath distortion. In this model, it is assumed that if a signalling scheme is "essentially bandlimited" to the frequency range $[-W, W]$ Hz then the channel is non-distorting. Section 3.2 presents some bandwidth measures and their interpretations while Section 5.4 links the fractional-power bandwidth to signal dimensionality. Consequently, the received electrical signal $y(t)$ can be written as

$$y(t) = x(t) + n(t), \tag{3.2}$$

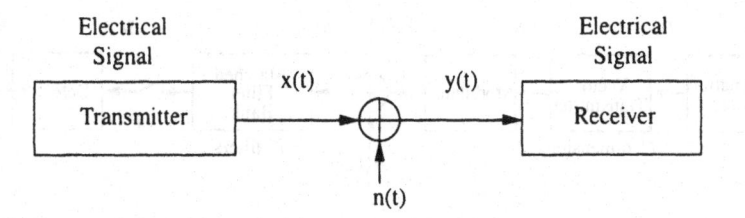

Figure 3.1. Communication system model for optical intensity channel in Figure 2.1.

which is reflected in the communications model in Figure 3.1.

The physical characteristics of the optical intensity channel impose constraints on the amplitude of $I(t)$ which can equivalently be viewed as constraints on $x(t)$. The amplitude non-negativity and average amplitude constraints of (2.1) and (2.2) can be cast on the electrical transmit signal as

$$(\forall t \in \mathbb{R}) \; x(t) \geq 0, \tag{3.3}$$

and

$$\lim_{T \to \infty} \frac{1}{2T} \int_{-T}^{T} x(t) \, dt \leq P. \tag{3.4}$$

To summarize, the wireless optical channel is modelled as a baseband electrical channel where signal $x(t)$ is transmitted so that $\forall t \; x(t) \geq 0$ and the average amplitude of $x(t)$ is upper bounded. The received electrical signal $y(t) = x(t) + n(t)$, where $n(t)$ is AWGN, can assume negative amplitude values.

Note that this channel model applies not only to free-space optical channels but also to fibre-optic links with negligible dispersion and signal independent, additive, white, Gaussian noise.

3.1.2 Vector Channel and Signal Space

In order to aid in the design and analysis of communication systems, the signals transmitted are often represented in a geometric framework as vectors in a linear space. This *signal space* is a convenient and compact means of representing the signals. The operation of the channel can also be viewed geometrically, and the impact of channel noise can be reflected in the same linear space. This *vector channel* models communication on the channel as the transmission of a vector and the reception of a corrupted vector in the same signal space. Figure 3.2 presents the channel model for an additive white Gaussian noise channel which models a wireless optical link corrupted by intense ambient illumination.

At the transmitter, an information source outputs a message at a rate of $1/T$ symbols per second. Without loss of generality, say that the message emitted by the information source is taken from the finite set $\mathsf{M} = \{1, 2, \ldots, M\}$. Each

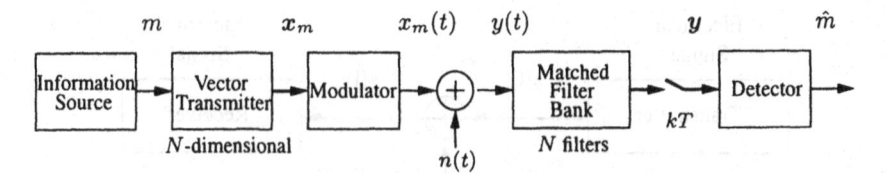

transmitted information symbol, is associated with a unique optical intensity signal, $x_m(t)$, which is transmitted on the channel in response to message m. Let $X = \{x_m(t) : m \in M\}$ be the set of transmitted optical intensity signals and let R_X be the common support of all functions in X.

In order to represent each of the transmitted signals in X geometrically, it is necessary to define a signal space model. A *signal space* corresponding to a set of optical intensity signals is an N-dimensional vector space, S, of real, square integrable functions of time with support set R_X, such that $X \subset S$ and $N \leq M$. Notice that the condition $N \leq M$ is satisfied with equality, if and only if the M elements of X are linearly independent. Define the index set $N = \{1, 2, \ldots, N\}$. Furthermore, define an inner product on the space S for functions $f(t), g(t) \in S$ as,

$$\langle f, g \rangle = \int_{R_X} f(t)g(t)dt.$$

Since S is a finite dimensional inner product space, there must exist a finite basis set of orthonormal functions in S which span the space. Let

$$\Phi = \{\phi_n(t) : n \in N\}$$

be a set of N real orthonormal function in S implying that $X \subset \text{span}(\Phi)$. Since the basis functions are orthonormal they satisfy the condition,

$$\langle \phi_k(t), \phi_l(t) \rangle = \begin{cases} 1 & : \quad k = l \\ 0 & : \quad k \neq l \end{cases}.$$

It is then possible to represent each $x_m(t) \in X$ as a linear combination of the elements of Φ as,

$$x_m(t) = \sum_{n \in N} x_{m,n} \phi_n(t),$$

for some real coordinates $x_{m,n}$. The vector of coordinates

$$x_m = (x_{m,1}, x_{m,2}, \ldots, x_{m,N})$$

with respect to the basis set Φ is termed a *signal vector* and can then be used to represent each transmitted signal waveform. The signal *constellation*, Ω, is defined as the collection of all the signal vectors,

$$\Omega = \{x_m : m \in \mathsf{M}\}.$$

Notice that the square norm of $f(t) \in \mathcal{S}$,

$$\langle f, f \rangle = \|f\|^2 = \int_{R_X} f(t)^2 dt$$

which is the electrical *energy* of the time signal. For each $x_m(t) \in \mathsf{X}$, the electrical energy of the signal can be written as,

$$\int_{R_X} x_m(t)^2 dt = \langle x_m, x_m \rangle$$
$$= \sum_{n \in \mathsf{N}} x_{m,n}^2$$
$$= \|x_m\|^2$$

since the basis set Φ is orthonormal. Thus, in the electrical energy of all transmitted signals is represented geometrically as the square norm of the signal vector.

Therefore, a *modulation scheme* can be described by the pair (Ω, Φ) which describes the set X of all possible emitted symbols. Unless otherwise stated, it often implicitly assumed that all symbols are chosen independently and equally likely over X.

3.1.3   Isolated Pulse Detection

Consider the transmission of a single isolated pulse from the set X. In Section 3.2 the case of multiple pulse transmission and bandwidth constraints is briefly discussed. At the receiver the received photocurrent signal $y(t)$ is,

$$y(t) = x(t) + n(t)$$

where $n(t)$ is a white Gaussian random process with power spectral density height σ^2. The channel is modelled as having a wide bandwidth, to simplify the comparison of various schemes, as is customary in other analyses of wireless optical modulation schemes [11, 2, 77]. The corrupted pulse at the receiver, $y(t)$, is detected using a bank of N matched filters with impulse response $h_n(t)$. Each *matched filter* has an impulse response "matched" to a basis function, namely,

$$h_n(t) = \phi_n(-t).$$

Assuming that the transmitter and receiver are synchronized, statistics on the transmitted symbol are found by sampling the output of the matched filters at the symbol rate to produce the vector

$$y = (y_1, y_2, \ldots, y_N).$$

It has been shown that this selection of filters is optimal in the sense of maximizing the sampled signal-to-noise ratio as well as providing *sufficient statistics* for the transmitted signal $x(t)$ [19, 45]. Alternatively, it is not difficult to show that the sampled output of the matched filters can be interpreted as the *projection* of received noisy signal, $y(t)$, onto the signal basis defined by Φ, i.e.,

$$y_n = \langle y(t), \phi_n \rangle .$$

By employing the signal space model, the channel from transmitted vector to received vector for the isolated pulse can be modelled as the vector channel

$$y = x_m + n \tag{3.5}$$

where n is a length N vector with independent Gaussian distributed components with zero mean and variance σ^2. An estimate of the sent symbol, $\hat{m}$, is formed based on the received vector y. The conditional distribution of y given m is then,

$$f_{y|m \text{ sent}} = \frac{1}{(2\pi\sigma^2)^{N/2}} \exp\left(-\frac{\|y - x_m\|^2}{2\sigma^2}\right), \tag{3.6}$$

The best decision rule to use, in the sense of minimizing the probability of symbol error, is the *maximum a posteriori* (MAP) detector [45]. In the case that the source selects the symbols with equal probability the MAP decision rule is a *maximum likelihood* (ML) detector which chooses the $\hat{m}$ to maximize the posterior probability in (3.6). Equivalently, the ML detector assigns $\hat{m} = m$, for the m which minimizes the Euclidean distance criterion $\|y - x_m\|^2$. In this manner, ML decoding is equivalent to minimum-distance decoding [19].

The signal space construction allows for the modelling of the communications channel as a vector channel and represents the electrical energy of every signal geometrically as the squared magnitude of the signal vector. Additionally, in the case of ML detection, the signal space provides a geometric interpretation of detection as minimum distance decoding. However, this signal space construction does not represent amplitude constraints nor the average amplitude of the optical intensity waveforms. Although the signal space model is appropriate for detection, since it is done in electrical domain, the channel constraints need to be represented in the model geometrically as well. Chapter 4 presents and extension to conventional electrical signal space models which allows for the representation of the optical intensity amplitude constraints.

3.1.4   Probability of Error

The estimation of the probability of a symbol error occurring in the detection process is a key parameter of any modulation scheme. The probability of symbol error is defined as,

$$P_{esym} = \Pr\{\hat{m} \neq m \mid m \text{ sent}\}.$$

In practice, this quantity is difficult to calculate exactly, and bounds are used to estimate its value. A common approximation on the probability of error is found using a *union bound* approximation [19, 45],

$$P_{esym} \approx \overline{N} \cdot Q\left(\frac{d_{min}}{2\sigma}\right), \qquad (3.7)$$

where d_{min} is the minimum Euclidean distance between two points in the signal constellation, $\overline{N}$ is the average number of constellation points d_{min} away from any point, and

$$Q(x) \triangleq \frac{1}{\sqrt{2\pi}} \int_x^\infty \exp(-u^2/2) \, du \qquad (3.8)$$

is the standardized Gaussian tail function. In the case of equally likely signalling, maximum likelihood detection is optimal in the sense of probability of symbol error and chooses the codeword with the minimum Euclidean distance to the received vector. At high signal-to-noise ratios, the impact of d_{min} on P_{esym} is much larger than that of $\overline{N}$ due to the exponential characteristics of the $Q(\cdot)$ function. Chapter 5 presents a more rigorous argument to allow for the comparison of optical intensity modulation schemes at high optical signal-to-noise ratios.

In the comparison of various modulation schemes, the probability of bit error, P_e, is a figure of merit. It is possible to determine P_e from P_{esym} exactly in some cases, while approximations must be made in others. A common approximation is that *Gray coding* is used between adjacent symbols in the constellation. A Gray coded constellation is labelled such that adjacent symbols differ by at most a single bit. At high signal-to-noise ratios, or equivalently at low P_{esym}, a symbol error is with high probability associated with a single bit error, i.e., it is most likely to make an error to a constellation point which is "closest" to the transmitted vector. Since there are $\log_2 M$ bits per symbol for an alphabet size of M, the approximation yields [19],

$$P_e \approx \frac{P_{esym}}{\log_2 M}.$$

3.1.5 Poisson Photon Counting Receiver

The Poisson photon counting channel models low intensity optical channels. In this regime, the main impediment to communication is due to the discrete nature of photons and electrons. As mentioned in Section 2.3, the random arrival times of photons, the random motion of electrons across the potential barrier include a random, yet signal dependent noise source in the link. In these types of links, the receiver is modelled as an electron counter over discrete intervals of time. If K is a random variable indicating the number of electrons received in a given interval, the probability of receiving k electrons is modelled as the Poisson distribution

$$\Pr\{K = k\} = \exp(-\lambda)\lambda^k/k!,$$

where λ is the average number of received electrons per interval. Although the Poisson channel model is more appropriate when applied to optical channels in which the receive intensity is low, such as fiber optical links, nonetheless it is an important related channel since it too suffers from similar amplitude constraints as does the indoor wireless optical channel. Section 3.3.4 presents some results on optimal signalling on Poisson channels.

3.1.6   Time-Disjoint Signalling

Consider the special case when $X = \{x_m(t) : m \in M\}$ is a set of optical intensity signals satisfying $(\forall m \in M)$ $x_m(t) = 0$ for $t \notin [0, T)$ for some positive symbol period T. Such an optical intensity set is termed *time-disjoint* since symbol waveforms shifted by integer multiples of T do not overlap in time. In the case where time-disjoint signals are sent independently, the optical intensity signal can be formed as

$$x(t) = \sum_{k=-\infty}^{+\infty} x_{A[k]}(t - kT), \tag{3.9}$$

where $A[k]$ is an i.i.d. process over M.

Since the symbols do not overlap in time, the non-negativity constraint (3.3) is equivalent to

$$(\forall m \in M, \ t \in [0, T)) \ x_m(t) \geq 0. \tag{3.10}$$

The average optical power calculation in (3.4) can also be simplified in the time-disjoint case as

$$\lim_{j \to \infty} \frac{1}{2(j+1)T} \sum_{k=-j}^{j} \int_0^T x_{A[k]}(t) \, dt \leq P,$$

which by the strong law of large numbers gives

$$P \geq \sum_{m \in M} \Pr(m) \left(\frac{1}{T} \int_0^T x_m(t) dt \right)$$

$$= E\left[\frac{1}{T} \int_0^T x_m(t) dt \right] \tag{3.11}$$

with probability one, where $\Pr(m)$ is the probability of transmitting $x_m(t)$. Thus, the average optical power, P, of a scheme is the expected value of the average amplitude of $x_m(t) \in X$.

3.2 Bandwidth

3.2.1 Definition

Since communication systems are often modelled as linear dispersion channels, the bandwidth of the channel is a key figure of merit of any channel. However, since the transmitted signals are modelled as being random, their Fourier transforms and spectral characteristics are also random. For a wide-sense stationary or wide-sense cyclostationary random process it is possible to define a quantity called the *power spectral density* (psd). The psd is an averaged second order statistic of a random process, and gives insight on the distribution of power over frequency.

As discussed in Chapter 2, the electrical characteristics of the channel impose the bandwidth penalty to the data bearing signal. Consider a modulation scheme (Ω, Φ) with transmitted waveform $x(t)$ which is modelled as a cyclostationary random process with power spectral density $S_X(f)$. The power spectral density consists of two components: a discrete spectrum, $S_X^d(f)$, and a continuous spectrum, $S_X^c(f)$ as

$$S_X(f) = S_X^d(f) + S_X^c(f). \tag{3.12}$$

The discrete spectrum consists of a series of Dirac delta, $\delta(\cdot)$, functions at various frequencies. Discrete spectral components are typically undesirable since they do not carry any information but require electrical energy to be transmitted. In the optical channel model, the discrete spectral component at $f = 0$, i.e., at DC, represents the average optical power of (Ω, Φ) while all other discrete components of the spectrum represent zero average optical power. The continuous spectrum is shaped by the distribution of the data symbols as well as by the pulse shape itself. Line coding can be used to introduce nulls in the spectrum, or to tailor the spectral response of the modulated signal to the channel in question.

For digital modulation schemes where the signal transmitted can be described as in (3.9) and the correlation from symbol to symbol can be described by a Markovian model, the discrete and continuous portions of the power spectral density take the form [78]

$$S_X^d(f) = \frac{1}{T^2} \sum_{n=-\infty}^{\infty} \left| \sum_{i \in M} \Pr(i) x_i^F \left(\frac{n}{T} \right) \right|^2 \delta \left(f - \frac{n}{T} \right),$$

and

$$S_X^c(f) = \frac{1}{T} \sum_{k \in M} \sum_{l \in M} \Pr(k) x_k^F(f) x_l^{F*}(f)$$

$$\left\{ \sum_{m=-\infty}^{\infty} \left(a_{kl}^{(m)} - \Pr(l) \right) \exp(-j2\pi f m T) \right\},$$

where $x_i^F(f)$ is the Fourier transform of signal $x_i(t) \in X$, $Pr(i)$ is the steady state probability of transmitting symbol $x_i(t)$ and $a_{kl}^{(m)}$ is the m-step conditional probability of transmitting symbol $x_l(t)$ given the current symbol is $x_k(t)$. The power spectral density depends on two factors : the pulse shapes, through the $x_i^F(f)$, and on the correlation between symbols. Under the conditions of independent and equally likely signalling, as in Chapter 5, the term a_{ij} can be simplified as

$$ a_{ij}^{(m)} = \begin{cases} 1/M & : \quad m \neq 0 \\ \delta_{ij} & : \quad m = 0 \end{cases}, $$

where δ_{ij} is the Kronecker delta function,

$$ \delta_{i,j} = \begin{cases} 1 & : \quad i = j \\ 0 & : \quad \text{otherwise} \end{cases}. \tag{3.13} $$

The power spectral density in (3.12) can be simplified to yield

$$ S_X(f) = \frac{1}{M^2 T^2} \sum_{n=-\infty}^{\infty} \left| \sum_{i \in M} x_i^F \left(\frac{n}{T} \right) \right|^2 \delta \left(f - \frac{n}{T} \right) $$

$$ + \frac{1}{T} \left[\sum_{i \in M} \frac{1}{M} \left| x_i^F(f) \right|^2 - \left| \sum_{i \in M} \frac{1}{M} x_i^F(f) \right|^2 \right]. \tag{3.14} $$

It should be noted that the electrical power of (Ω, Φ),

$$ \int_{-\infty}^{\infty} S_X(f) \, df $$

is *not* equivalent to the optical power cost of the scheme. The power spectrum is the distribution of electrical energy in the transmitted signal, $x(t)$ in Figure 3.1, while the average optical power is the average signal amplitude. As discussed in Section 3.1.1, the optical channel can be modelled as a baseband electrical system with constraints on the amplitude of signals transmitted. As a result, the use of the bandwidth of the electrical signals is appropriate.

The bandwidth occupied by a modulation scheme is a measure of the amount of spectral support needed for the transmission of the signal. Since all real signals have even magnitude spectra, only positive frequencies are considered. However, in systems employing time-disjoint symbols or in general non-bandlimited signals, the definition of the bandwidth of (Ω, Φ) is non-trivial. There are several definitions of bandwidth which are popular in literature. In the electronics domain, the -3 dB bandwidth occurs at the frequency when the psd is a factor $1/\sqrt{2}$ lower than the peak value. The first-null bandwidth is defined as the width in positive frequencies of the main lobe of the signal.

Although most of the energy of the pulse is contained in the main lobe, this measure does not penalize schemes with large side-lobe power. This bandwidth measure has been used in many previous reference on optical intensity modulation for wireless channels [79, 11, 2, 77]. This spectral measure is misleading in the case of passband signals and generally useful only for lowpass signals.

The *fractional power bandwidth* is a superior measure of signal bandwidth since it defines the signal bandwidth as the extent of frequencies where the majority of the signal power is contained, as opposed to arbitrarily denoting the position of a particular spectral feature as the bandwidth. Define the fractional power bandwidth, W_K, as satisfying the expression,

$$\frac{\int_{-W_K}^{W_K} S_X^c(f)\, df}{\int_{-\infty}^{+\infty} S_X^c(f)\, df} = K, \qquad (3.15)$$

where $K \in (0, 1)$ is fixed to some value typically 0.99 or 0.999. The channel model of Section 3.1.1 assumes that the frequency response of the channel is flat and that signals are limited to a bandwidth of $[-W_K, W_K]$. To a first approximation, if K is chosen large enough, the energy outside of this band lies below the noise floor of the channel and neglecting it introduces little error. In this sense, (Ω, Φ) is considered as being "essentially" bandlimited to the channel bandwidth.

For example, for rectangular on-off keying, described in Section 3.3.1, with symbol interval T and average optical power P, the power spectral density is

$$S_X(f) = \delta(f) + T\text{sinc}^2(\pi f T),$$

where

$$\text{sinc}(x) \triangleq \frac{\sin(x)}{x}. \qquad (3.16)$$

Notice that the discrete component at DC represents the average optical power cost of the modulation scheme. If the continuous portion of the psd is used to define the bandwidth, the -3 dB bandwidth is $f \approx 0.44/T$, the 90% fractional bandwidth is $f \approx 0.9/T$, and the first-null bandwidth is $f = 1/T$.

The *bandwidth efficiency* of a modulation scheme is given as the ratio of the bit rate R in bits/second and bandwidth B in Hz, and is used as a figure of merit in the comparison of modulation schemes. Chapter 5 discusses the use of this metric and the relation to the dimension of the signal space.

The bandwidth of a signal can also be interpreted as a measure of the number of degrees of freedom available or number of *dimensions* available. For example, in the case of ideally bandlimited signals to bandwidth $W = 1/2T$ Hz, the channel permits a single degree of freedom, or dimension, every T seconds. The *Landau-Pollak* dimension [80] of the set of signals formalizes and extends this concept to cast a spectral constraint as an effective number of dimensions.

Let $L^2[0, T]$ denote the set of all finite energy signals with support contained in $[0, T)$. Define the $(1 - \epsilon)$-fractional energy bandwidth, $W_\epsilon(x)$, of a transmitted symbol $x(t) \in L^2[0, T]$ with Fourier transform $X(f)$ as

$$W_\epsilon(x) = \inf \left\{ W \in [0, \infty) : \int_{-W}^{W} |X(f)|^2 df \geq (1 - \epsilon) \int_{-\infty}^{\infty} |X(f)|^2 df \right\}$$

(3.17)

where $\epsilon \in (0, 1)$ is fixed to some value, typically 10^{-2} or 10^{-3}. This bandwidth measure quantifies the frequency concentration of $x(t)$. In practical terms, $x(t)$ can be thought of as being effectively band-limited to $W_\epsilon(x)$ Hz if ϵ is chosen so that the out-of-band energy is below the noise floor of the channel [81].

A dimension for the signal $x(t) \in L^2[0, T]$ can be defined through the fractional energy bandwidth. Consider approximating $x(t) \in L^2[0, T]$ as a linear combination of some orthonormal basis functions. For a given $W_\epsilon(x)$ and T, the best such basis, in the sense of minimizing the energy in the error of the approximation, is the family of prolate spheroidal wave functions, $\varphi_n(f)$ [82]. The $\varphi_n(f)$ are functions strictly time-limited to $[0, T)$ which have the maximum energy in $[-W_\epsilon(x), W_\epsilon(x)]$ of all unit energy functions [83]. The error in the approximation can be upper bounded as [80]

$$\inf_{\{a_i\}} \int_{-\infty}^{\infty} \left| X(f) - \sum_{n=0}^{\lceil 2W_\epsilon(x)T \rceil} a_n \varphi_n(f) \right|^2 df < 12\epsilon^2.$$

(3.18)

In this sense the signal $x(t)$ can be thought of as being indistinguishable from some linear combination of prolate spheroidal basis functions. It can then be said that $x(t)$ is essentially $2W_\epsilon(x)T$ dimensional with the error in the approximation tending to zero as $\epsilon \to 0$. This definition will be used in Chapter 6 to allow for the interpretation of channel capacity results in terms of the channel bandwidth.

3.2.2   Inter-Symbol Interference

In the case of isolated pulse transmission, treated in Section 3.1.3, no consideration is given to the more realistic case of transmitting a train of pulse symbols. In this case, the interaction between successive transmitted pulses must be taken into consideration. In dispersive channels, transmitted pulses are received, after matched filtering, with some *inter-symbol interference* (ISI). That is, the sampled output of the matched filter depends not only on the symbol sent at that time, but, in general, on *all* previous and future transmitted symbols. Say that $p(t)$ is the signal at the output of the matched filter in Figure 3.2, the *Nyquist condition* for zero ISI is that the sampled output of the matched filter satisfies,

$$p(kT) = \delta_{k,0},$$

where $\delta_{k,0}$ is the Kronecker delta function (3.13). This condition can be interpreted as ensuring that at a given sample time the output of the matched filter depends on a single transmitted symbol. The zero-ISI condition can be written in frequency domain as

$$\sum_{m=-\infty}^{\infty} P\left(\omega - m\frac{2\pi}{T}\right) = T.$$

The minimum bandwidth *Nyquist pulse*, which satisfies the zero-ISI condition for bandwidth W is $\mathrm{sinc}(2\pi W t)$, while the minimum bandwidth Nyquist pulse subject to an amplitude non-negativity constraint is $\mathrm{sinc}^2(\pi W t)$ [5]. On low pass channels, such as some wireless optical channels, the impact of ISI can be reduced through the use of transmitted signals which have a significant proportion of their energy in the passband of the channel.

In order to detect the transmitted signals a maximum likelihood sequence detector (MLSE) is required, assuming all sequences are equally likely. In general the complexity of such detectors is prohibitively high and a lower complexity filtering process known as *equalization* is typically done. An equalizer is a filter, either linear or non-linear, which mitigates the impact of ISI on the performance of the system. Typically, a minimum mean-square error criterion is used in the design of equalizers instead of the zero-ISI condition to reduce undue noise enhancement. Section 3.3.4 presents some references on equalization on wireless optical channels especially for schemes such rectangular OOK and PPM which are optical power efficient but bandwidth inefficient.

This section has presented some of the key terminology, however, many excellent references exist on the design and analysis of signalling and bandwidth limited channels [19, 45, 84].

3.3 Example Modulation

Conventional wireless optical systems have been conceived using a variety of binary and multi-level modulation formats. This section briefly overviews several important multi-level modulation schemes in wireless optical channels. The framework for analysis mirrors that used in previous work on signalling design for wireless optical channels and is included for completeness [2].

3.3.1 Binary-Level Modulation

Most popular schemes in use on wired and wireless optical communication systems rely on binary level modulation schemes. These modulation techniques transmit information in each symbol period through the variation of two intensity levels. An advantage of these schemes is that they typically have simple and inexpensive implementations. This section presents an analysis of two well known modulation schemes on optical links, *on-off keying* (OOK) and *pulse position modulation* (PPM).

On-Off Keying

On-Off keying is a popular modulation scheme not only in wireless infrared links, but also in a wide variety of data communication applications. In many conventional channels, this scheme is also known as non-return-to-zero (NRZ) encoding.

On-Off keying is a binary level modulation scheme consisting of two symbols. In each symbol interval one of the two symbols is chosen with equal probability. The transmitted symbols consist of constant intensities of zero or $2P$ through the symbol time. The signal can be represented by the basis function for OOK, $\phi_{OOK}(t)$, illustrated in Figure 3.3. This basis function is defined as

$$\phi_{OOK}(t) = \frac{1}{\sqrt{T}} \text{rect}\left(\frac{t}{T}\right), \tag{3.19}$$

where

$$\text{rect}(t) = \begin{cases} 1 & : & 0 \leq t < 1 \\ 0 & : & \text{otherwise.} \end{cases}$$

Using this basis function, an expression for the time varying optical intensity is

$$x(t) = \sum_{k=-\infty}^{\infty} 2P\sqrt{T}A[k]\phi_{OOK}(t - kT) \tag{3.20}$$

where $A[k] \in \{0, 1\}$ and chosen uniformly. Since the basis function is non-negative in the symbol period, and since only non-negative multipliers are used, $x(t)$ satisfies the non-negativity constraint. The average amplitude of $x(t)$ is set at P due to the distribution of the data symbols and the scaling of the multipliers.

The constellation for OOK consists of two points in a one dimensional space as illustrated in Figure 3.3. The probability of a bit error can be determined using the previously defined framework as

$$P_e = Q\left(\frac{P}{\sqrt{R\sigma^2}}\right)$$

since the bit rate $R = 1/T$ and $P_{esym} = P_e$ in this case.

The power spectral density of OOK was calculated to determine the bandwidth of the system. Using (3.14), the power spectral density of OOK was found to be

$$S_{OOK}(f) = P^2\delta(f) + P^2T \text{sinc}^2(\pi fT),$$

where $\text{sinc}(\cdot)$ is defined in (3.16). Figure 3.4 contains a plot of the continuous portion of power spectral density of OOK. It is clear that the first-null band-

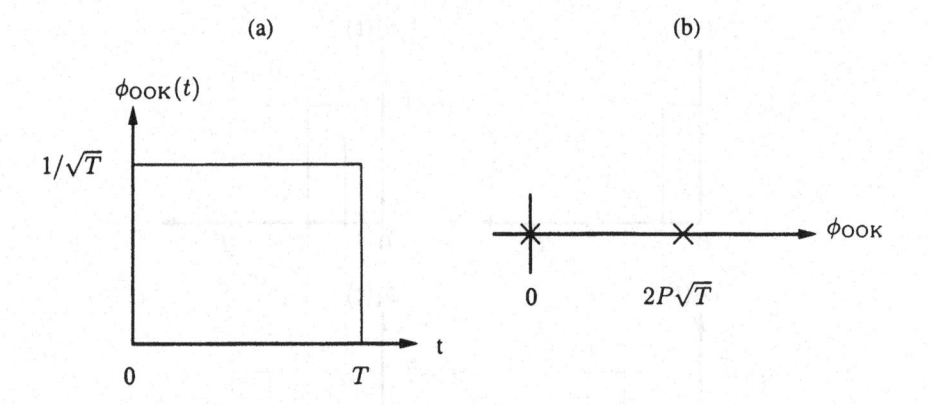

Figure 3.3. Basis function (a) and constellation (b) of on-off keying.

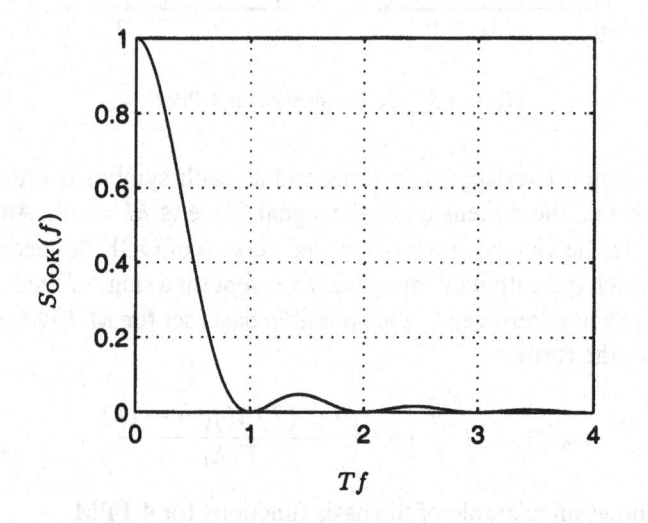

Figure 3.4. The continuous portion of the power spectral density of on-off keying, for $P = 1$ and $T = 1$.

width, as defined earlier, is $1/T$ Hz. In this bandwidth measure, the bandwidth efficiency of this scheme is thus $R/B = 1$ bit/s/Hz.

Pulse-Position Modulation

Pulse position modulation (PPM) is a standard modulation technique used in optical communications. The IrDA specification for the 4 Mbps short distance wireless infrared link specifies a 4-PPM modulation scheme [52, 85].

The family of M-PPM modulation schemes are M-ary modulation schemes using two distinct intensity levels. Each symbol interval is divided into a series of M subintervals, or chips. Information is sent by transmitting a non-zero optical intensity in a single chip, while other chip intervals remain dark. Each

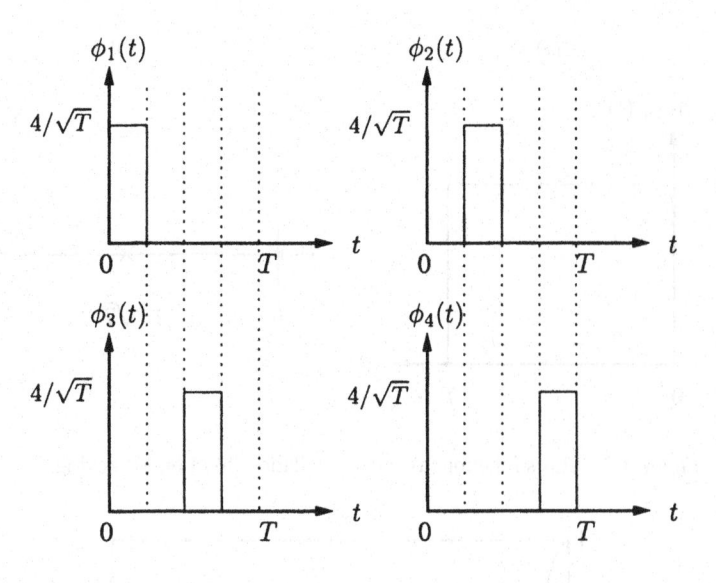

Figure 3.5. Basis functions for 4-PPM.

of the chips is non-overlapping in time, and so each symbol is orthogonal to all the others, i.e., the dimension of the signal space is $M = N$. An M-PPM symbol can also be viewed as a block coded version of OOK defined over MT seconds in which the output intensity is zero except for a single T interval which assumes a fixed non-zero value. One possible basis set for M-PPM, $\phi_m(t)$ for $m \in \mathsf{M}$, takes the form

$$\phi_m(t) = \sqrt{\frac{M}{T}} \operatorname{rect}\left(\frac{t - (T/M)(m-1)}{T/M}\right). \qquad (3.21)$$

Figure 3.5 shows an example of the basis functions for 4-PPM.

The signal space of M-PPM is an M dimensional Euclidean space with a single constellation point on each of the M axes. A time domain representation of the intensity waveform as sent on the channel is

$$x(t) = \sum_{k=-\infty}^{\infty} MP\sqrt{\frac{T}{M}} \phi_{A[k]}(t - kT),$$

where $A[k]$ chooses the symbol equiprobably in M. The pulses remain non-negative for all time due to their construction. The average optical power of each symbol is fixed at P by setting the peak value of each symbol to MP. The information in this system is transmitted in the position of the pulse within the symbol interval.

The probability of error for this modulation scheme can be calculated by noting that each constellation point is orthogonal to all others and that each constellation point is equidistant to all the other points. Based on this geometry the probability of symbol error can be found as

$$P_{esym} \approx (M - 1) \cdot Q \left(P \sqrt{\frac{M}{2R_s \sigma^2}} \right)$$

where $R_s = 1/T$ is the symbol rate. Due to the orthogonality of all the points in the space, and the fact they are equiprobable, the probability of symbol error can be converted to a probability of bit error by multiplying by a factor of $\frac{M}{2}/(M-1)$ [84]. Combining these results gives the probability of bit error as

$$P_e \approx \frac{M}{2} \cdot Q \left(P \sqrt{\frac{M \log_2 M}{2R\sigma^2}} \right)$$

since the bit rate $R = R_s \log_2 M$ in this case. Using (3.14), the power spectral density of M-PPM is

$$S_{PPM}(f) = P^2 \delta(f)$$
$$+ P^2 T \operatorname{sinc}^2 \left(\frac{\pi f T}{M} \right) \left[1 - \frac{1}{M^2} \left(M + 2 \sum_{i=1}^{M-1} (M - i) \cos \left(\frac{2\pi f T}{M} i \right) \right) \right].$$

Figure 3.6 contains an example of the continuous power spectral density for 4-PPM. This result can be generalized to show that the occupied first-null bandwidth is $B = M/T = MR_s$. Following this expression, the bandwidth efficiency with the given bandwidth definition is given as

$$\frac{R}{B} = \frac{1}{M} \log_2 M \quad \frac{\text{bits/s}}{\text{Hz}}$$

since $R_s = R/\log_2 M$.

3.3.2 Multi-Level Modulation

Traditional approaches to modulation for the optical intensity channel center on developing power efficient schemes. The bandwidth of the transmitted symbols is not a critical issue due to the wide bandwidth available in fibre systems. However, the wireless optical channel relies on the use of inexpensive devices which are severely limited in available bandwidth, as shown in Chapter 2. As a result, for wireless optical links, the bandwidth efficiency of the modulation scheme is a parameter of critical importance.

Multilevel modulation techniques transmit symbols in which the intensity values are continuous in a range or take on a set of values. The advantage of these schemes is that they provide a higher bandwidth efficiency than binary

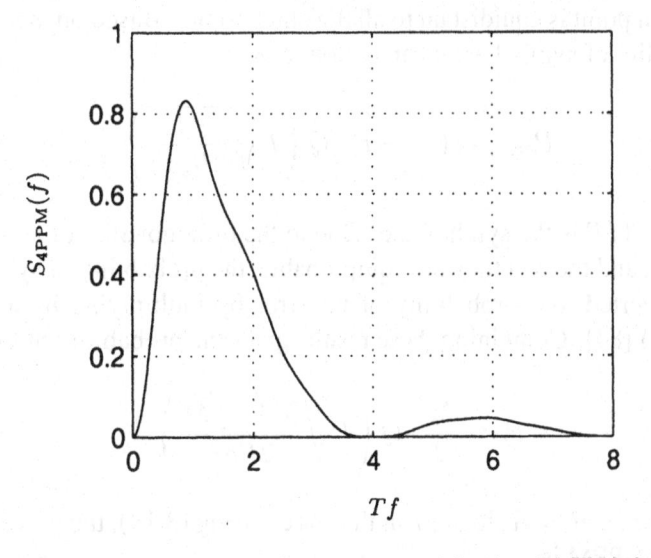

Figure 3.6. The continuous portion of the power spectral density of 4-PPM, for $P = 1$ and $T = 1$.

level techniques, since data is transmitted in the amplitude level as well as the structure of the basis functions.

This section presents a brief analysis of two classical multilevel schemes for the intensity modulated channel: *pulse amplitude modulation* (PAM) and *quadrature pulse amplitude modulation* (QAM).

Pulse Amplitude Modulation

Pulse amplitude modulation is a classical modulation scheme which can be adapted to operate on an intensity modulated optical channel. This scheme is a generalization of on-off keying from a set of two symbols to a set of M symbols. The basis function for rectangular M-PAM, $\phi_{PAM}(t) = \phi_{OOK}(t)$ as given in (3.19). In fact, any non-negative pulse time-disjoint pulse can be chosen as a basis function. The basis function for rectangular PAM and constellation are illustrated in Figure 3.7.

In each symbol period one of M scaling factors are chosen with equal probability. These scaling factors are chosen so that for an equiprobable distribution of transmitted symbols, the average output amplitude is fixed at P. Unlike PPM, where the average of the modulation scheme is independent of the symbols chosen, PAM requires prior knowledge of the data distribution in order to ensure the average optical power is limited. The non-negativity constraint of the optical intensity channel is met by choosing M non-negative scaling factors. The time varying intensity signal corresponding to PAM which satisfies

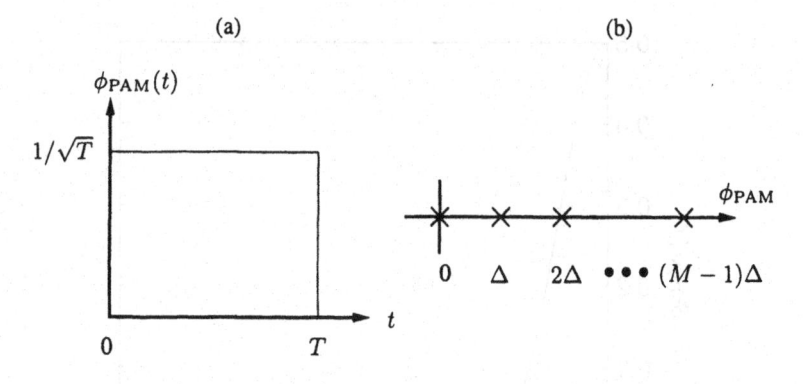

Figure 3.7. Basis function (a) and constellation (b) of pulse amplitude modulation (where $\Delta = 2\sqrt{T}P/L - 1$).

the channel constraints is

$$x(t) = \sum_{k=-\infty}^{\infty} \frac{2P}{M-1}\sqrt{T}(A[k] - 1)\phi_{\text{PAM}}(t - kT)$$

where $A[k]$ is uniform over M. Note that for $M = 2$ this expression is the same as the intensity waveform specified in (3.20) for OOK. As a result, OOK is a special case of rectangular PAM.

The probability of symbol error can be determined by constructing the constellation for this modulation scheme. Since all signal points can be expressed in terms of one basis function, the constellation is one dimensional. The M constellation points are evenly spaced on the axis as is illustrated in Figure 3.7. The probability of symbol error can be determined by assuming each symbol is sent with equal probability as

$$P_{esym} = \frac{2}{M}(M-1) \cdot Q\left(\frac{P}{M-1}\sqrt{\frac{1}{R_s\sigma^2}}\right).$$

Assuming Gray coding is used, the probability of bit error is approximately

$$P_e \approx \frac{2}{M \log_2 M}(M-1) \cdot Q\left(\frac{P}{M-1}\sqrt{\frac{\log_2 M}{R\sigma^2}}\right)$$

since the bit rate $R = R_s \log_2 L$.

The power spectrum of PAM can be calculated using (3.14). Assuming equiprobable symbol distribution the psd for M-PAM is

$$S_{\text{PAM}}(f) = P^2\delta(f) + P^2T\frac{M+1}{3(M-1)}\operatorname{sinc}^2(\pi fT).$$

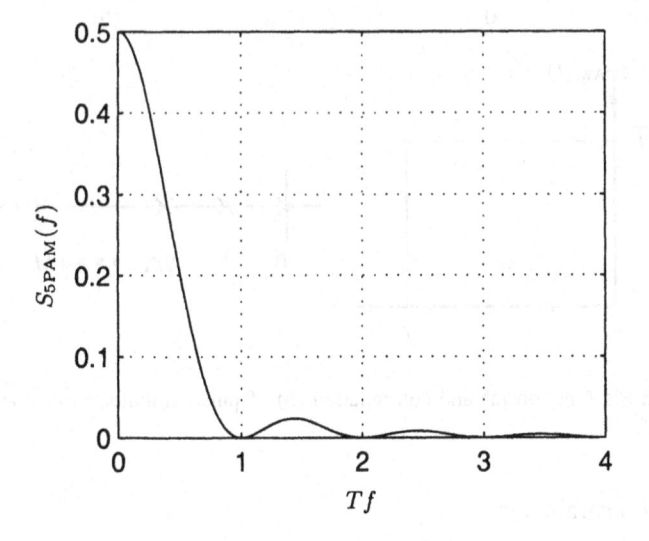

Figure 3.8. The continuous portion of the power spectral density of 5-PAM, for $P = 1$ and $T = 1$.

The continuous portion of the power spectral density of rectangular 5-PAM is plotted in Figure 3.8. As is the case with all modulation schemes for the optical intensity channel, the DC component is fixed at P. The first-null bandwidth of this scheme is $B = R_s$, as was the case in OOK. The addition of multiple levels per symbol allows for more than 1 bit of information per transmitted symbol. The bandwidth efficiency of PAM on the intensity modulated channel is

$$\frac{R}{B} = \log_2 M \ \frac{\text{bits/s}}{\text{Hz}}. \tag{3.22}$$

Quadrature Pulse Amplitude Modulation

Quadrature pulse amplitude modulation (QAM) is a popular modulation scheme in conventional channels such as high speed wired channels and radio frequency radio channels. The basic structure of QAM is unaltered in the wireless optical channel, however, changes must be made to ensure the channel constraints are met.

The M^2 symbols of M^2-QAM consist of an in-phase and quadrature component basis function which are orthogonal to each other, as shown in Figure 3.9. The basis functions, $\phi_I(t)$ and $\phi_Q(t)$, represent the in-phase and quadrature components of the data signal and take the form

$$\phi_I(t) = \sqrt{\frac{2}{T}} \cos\left(\frac{2\pi}{T} qt\right) \text{rect}\left(\frac{t}{T}\right)$$

$$\phi_Q(t) = \sqrt{\frac{2}{T}} \sin\left(\frac{2\pi}{T} qt\right) \text{rect}\left(\frac{t}{T}\right). \qquad (3.23)$$

for some integer $q \geq 1$. In each symbol instant, independent data is used to modulate the two basis functions. As a result, each basis function is multiplied by a series of M amplitude values to comprise the M^2 symbols. Taking into account the channel constraints, the transmitted signal is

$$x(t) = \sum_{k=-\infty}^{\infty} \frac{\sqrt{T}P}{2(M-1)} \left(A[k]\phi_I(t-kT) + B[k]\phi_Q(t-kT)\right)$$

$$+ P \cdot \text{rect}\left(\frac{t-kT}{T}\right). \qquad (3.24)$$

for $A[k], B[k] \in \{-(M-1), -(M-3), \ldots, (M-1)\}$. Since the basis functions $\phi_I(t)$ and $\phi_Q(t)$ take on negative amplitudes at intervals within the symbol period, a DC bias must be added to ensure the non-negativity constraint is met. Conventional QAM adds a fixed bias of P to every symbol. This added DC bias is rejected by the matched filter front end since the basis functions have an average value of zero. The addition of a fixed DC bias also ensures that the average power constraint in (3.4) is met, independent of the data distribution. The maximum amplitude value for the data is fixed at $P\sqrt{T}/2$, and the minimum amplitude value for these symbols is 0. However, for symbols with smaller swing values the minimum value of the symbol is greater than zero. This is the condition of excess symbol power. Chapters 4 and 5 describe techniques in which the symbol bias varies *per symbol* in order to improve the optical power efficiency of QAM modulation.

The constellation of QAM is a two dimensional regular array of points, as illustrated in Figure 3.9. The minimum spacing between the points is fixed by the amount of DC bias added due to the non-negativity constraint. It can be shown from (3.24), that

$$d_{\min} = \frac{P}{M-1} \sqrt{\frac{2 \log_2 M}{R}}.$$

The probability of symbol error can be approximated by applying the union bound approximation in (3.7). By computing the average number of neighbours $d_{\min}$ away from each constellation point, an estimate for P_{esym} can be formed as

$$P_{\text{esym}} \approx \frac{4(M-1)}{M} \cdot Q\left(\frac{P}{M-1} \sqrt{\frac{1}{4R_s\sigma^2}}\right).$$

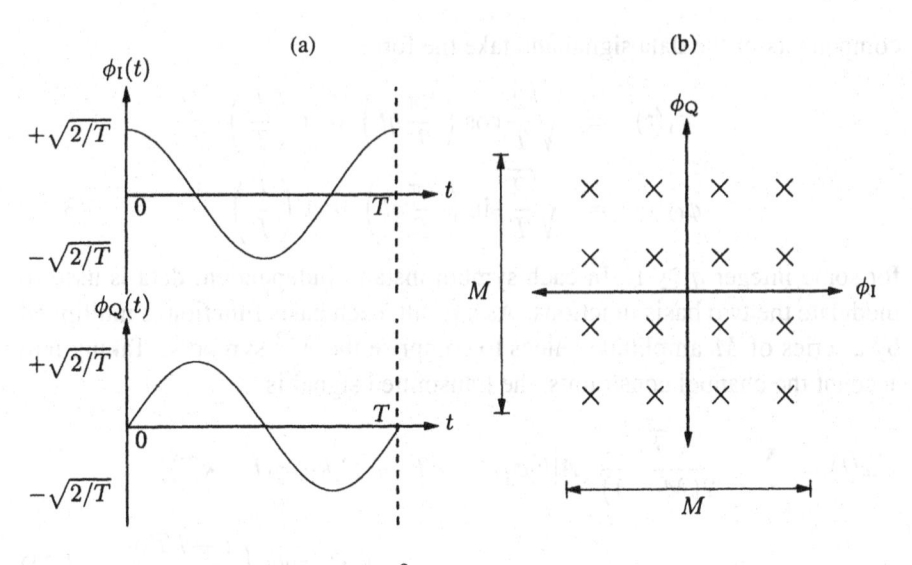

Figure 3.9. Basis functions (a) and $M^2 = 16$ point constellation (b) of quadrature pulse amplitude modulation (for $q = 1$).

Using the Gray coding approximation, the probability of bit error for M^2-QAM is

$$P_e \approx \frac{2(M-1)}{M \log_2 M} \cdot Q\left(\frac{P}{M-1}\sqrt{\frac{\log_2 M}{2R\sigma^2}}\right)$$

where $R = R_s \log_2 M^2$.

The spectral characteristics of QAM can be investigated by computing the power spectral density via (3.14). The power spectral density for the L^2-QAM signal in (3.24) is

$$S_{QAM}(f) = P^2\delta(f)$$
$$+ P^2T\frac{M+1}{12(M-1)}\left[\text{sinc}^2(\pi Tf - \pi) + \text{sinc}^2(\pi Tf + \pi)\right].$$

Figure 3.10 illustrates an example of the continuous portion of the psd for 25-QAM. Since QAM is a passband modulation scheme with little signal energy at low frequencies it is robust in environments in which low frequency interferers are dominant, such as the noise created by fluorescent lighting ballasts discussed in Section 2.3.

The first-null bandwidth of M^2-QAM, as defined by the frequency of the first spectral null, is $B = 2R_s$ Hz. Following this result, the bandwidth efficiency of M^2-QAM is

$$\frac{R}{B} = \log_2 M \quad \frac{\text{bits/s}}{\text{Hz}},$$

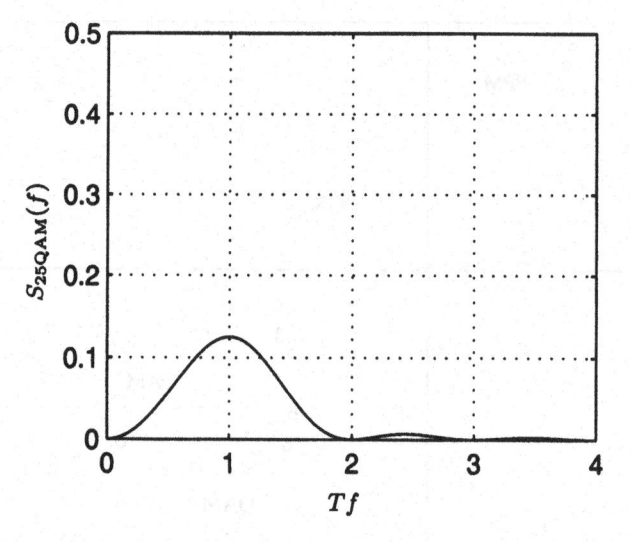

Figure 3.10. The continuous portion of the power spectral density of 25-QAM, for $P = 1$ and $T = 1$.

which is identical to the bandwidth efficiency of M-PAM in (3.22). Although the bandwidth required for M^2-QAM is double that of M-PAM, the amount of information transmitted by the QAM symbols is double that of PAM, which satisfies intuition.

Notice that for different values of q in (3.23) that the resulting in-phase and quadrature basis functions are orthonormal. A modulation scheme consisting of the sum of QAM signals, each at different values of q, transmits information by multiplexing in frequency domain. This type of modulation is termed *multiple sub-carrier modulation* (MSM). The advantage of MSM is due to the ability to multiplex many independent data streams in a single transmission. The impact of frequency selective channels can also be mitigated through the use of many narrow sub-carriers. Additionally, MSM shares the same benefit as QAM in the ability to reject low frequency cyclostationary noise generated by fluorescent light fixtures. However, it has been shown that the optical power efficiency of MSM schemes is far lower than for OOK or PPM modulation [2, 11, 86]. Section 3.3.4 discusses some techniques which have been developed to improve the average optical power efficiency of MSM techniques.

3.3.3 Discussion

In order to compare the modulation schemes discussed in this section, it is necessary to consider both the optical power efficiency of each scheme and the bandwidth efficiency. The optical power efficiency is measured by computing the optical power gain over OOK if the $d_{\min}$ of both schemes is equal. This is an

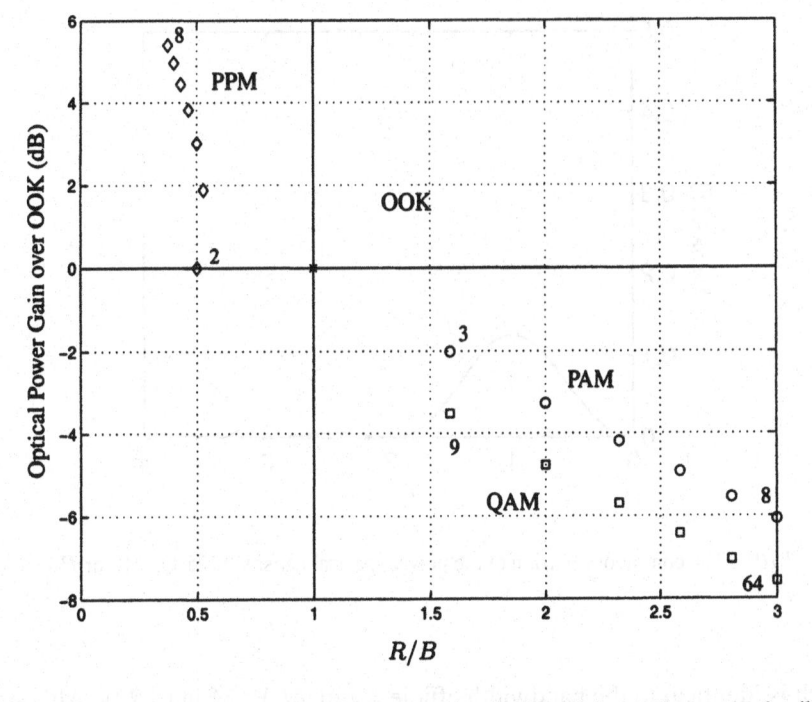

Figure 3.11. Optical power gain over OOK versus bandwidth efficiency (first spectral null) for conventional modulation schemes.

approximation of the relative optical power required by a given scheme versus OOK. Section 5.2 formalizes this comparison technique and demonstrates that it is the ratio of the constellation figure of merit of each scheme. The optical power gain is plotted versus the bandwidth efficiency, as measured by the first-null bandwidth. Figure 3.11 plots the comparison of optical power gain versus bandwidth efficiency for the schemes discussed in this section, however, more comprehensive comparisons using the first-null bandwidth as a metric have been reported [2].

The optimal point for any modulation scheme destined for wireless optical links would be in the upper right hand corner of the figure, where bandwidth efficiency and power efficiency are maximized. Binary level techniques offer the best power efficiency at the cost of reduced bandwidth efficiency. Multilevel schemes provide the necessary bandwidth efficiency for wireless optical links at the cost of power efficiency.

The characteristics of PPM are well suited to channels in which optical power, and not bandwidth, efficiency are of primary importance such as in fibre applications. Qualitatively, for a given average optical power, narrow pulse techniques trade-off pulse width for increased *electrical* energy, which improves detection. Optical power efficient PPM enjoys wide application in wireless and wired optical networks due to its high power efficiency. However, the price paid for this power efficiency is a reduction in the bandwidth efficiency of the scheme. Binary level schemes, although power efficient, do not provide a high degree of bandwidth efficiency. However, multilevel schemes trade-off power efficiency for bandwidth efficiency.

3.3.4 Previous Work

Although, studies into signalling over optical Poisson channels has been active for some time [87, 88], the field of indoor wireless optical channels has been an area of intense study since the landmark paper of Gfeller and Bapst [1].

Binary-level signalling is the simplest and most common modulation for optical intensity channels. Under the conditions of high background illumination and a constraint on the average optical power, it was shown in [87] that the average-distance-maximizing binary scheme on the Poisson channel is 2-PPM [87]. Later, the requirement for high background illumination was removed and such pulses were shown to be optimal in the same sense on an optical Poisson counting channel [88]. This result was confirmed for the case of the Gaussian noise channel in the case of maximum likelihood detection in [5]. For M-ary signalling, Gagliardi and Karp [89, 43] showed that pulse-position modulation (PPM) is optimal in the same average distance measure, on a Poisson counting channel for $M > 2$, but stopped short of declaring M-PPM as optimal for the optical counting channel.

The most prominent modulation formats for wireless optical links are binary level PPM and on-off keying (OOK). Low cost, point-to-point modems conforming to the IrDA standard utilize a 4-PPM scheme [85]. Due to the bandwidth-restricted nature of the wireless optical channel and some guided-wave optical systems, several variations of PPM have been developed; namely, multiple PPM, overlapping PPM, differential PPM, pulse interval modulation and edge position modulation. However, all remain binary level [90–92, 77, 11, 2, 93–95]. Due to the bandwidth inefficiency of high order PPM pulses and rectangular OOK, coded modulation and equalization have also been extensively investigated on Poisson and wireless optical channels [96–102, 23]. These schemes are investigated to reduce the impact of ISI on signal detection as well as to provide practical techniques to mitigate this interference, especially on diffuse channels.

Multi-level signalling formats have been proposed for wireless optical and photon counting channels [2, 103]. These schemes, in general, provide high spectral efficiencies at the expense of a loss in power efficiency. Multi-level *adaptively biased QAM* (AB-QAM) transmits data not only in the in-phase and quadrature carriers but also in the DC bias of each symbol [104, 5]. The bias of each AB-QAM symbol is minimized to improve optical power efficiency and the received DC bias value is used to provide a degree of diversity at the receiver, improving error performance. Filtered modulation using Gaussian pulse shapes for PPM symbols has also been proposed [105]. The use of sub-carrier modulation has also been investigated for this channel as a bandwidth efficient scheme to mitigate the effects of narrow band interferers at the expense of power efficiency [86, 2, 11]. Average optical power reduction techniques for multiple sub-carrier modulation have also been devised as a extension of the AB-QAM biasing concept [106].

On the fibre-optic channel, multi-level signalling in the presence of channel non-linearities was investigated in detail and shown to increase the link distance of 10 Gb/s systems [107].

3.4　　The Communication System Design Problem

The wireless optical channel presents unique and exciting possibilities for signalling and modem design. Chapter 2 outlined some of the physical properties of point-to-point, diffuse and quasi-diffuse channel topologies. In this chapter, the issues involved with signalling design have been introduced and some popular signalling schemes have been discussed.

The wireless optical intensity channel imposes amplitude constraints on all transmitted signals. Specifically, all emitted signals must be non-negative since optical intensity modulators and direct detection receivers are employed. Additionally, the average amplitude must also be constrained due to eye and skin safety requirements. These amplitude constraints lie in contrast to the constraints on mean-square amplitude constraints in conventional electrical channels. As a result, it is often not efficient or even possible to apply electrical signalling design techniques directly to the optical intensity channel.

An additional constraint shared by both optical intensity and electrical channels are bandwidth limitation. In the case of wireless optical channels these bandwidth limitations arise due to optoelectronic response limitations as well as due to multipath dispersion in diffuse links. In many popular modulation techniques for wireless optical channels, optical power efficiency comes at the price of low bandwidth efficiencies. Conversely, conventional bandwidth efficient signalling improve potential data rates at the expense of power efficiency.

The communications design problem is then one of carefully designing a pulse set in order to meet required amplitude constraint while at the same time ensuring that power and bandwidth efficiency targets are met. The conventional approach of adding a constant bias may lead to simple signalling, however, it can also lead to significant loss in power efficiency. The use of narrow rectangular pulse techniques, such as in PPM, yields large optical power gains, however, requires the use of equalization techniques to overcome the limitation of the bandlimited channel.

The following chapters discuss some solutions for the design of optical intensity signalling sets which considers the non-negativity, average optical power and bandwidth constraints. The following chapters address the signal design question by addressing how to represent optical intensity signals, how to design modulation and coding and what the ultimate gain for a given pulse set is. Chapter 4 presents a signal space model for optical intensity sets which represents non-negativity and average optical power constraints geometrically. Chapter 5 presents a method to design optical intensity signalling sets and gives a means to compare modulation schemes. Chapter 6 then addresses the fundamental limits on communications subject to the amplitude and bandwidth constraints.

PART II

SIGNALLING DESIGN

Chapter 4

OPTICAL INTENSITY SIGNAL SPACE MODEL

In electrical modem design, a signal space model is conventionally defined to represent the set of transmitted waveforms geometrically [19]. Each transmitted waveform is represented as a linear combination of the elements of some orthonormal basis set as described in Section 3.1.2. This is a powerful representation since the energy cost of each transmitted waveform is represented as the square norm of each signal vector. Additionally, for additive white Gaussian noise channels, the detection process acquires a rich geometric meaning. However, this construction does not represent amplitude constraints nor the average amplitude of the set of transmitted waveforms. The model is appropriate since detection is done in electrical domain.

This chapter presents a formal mathematical model for the optical intensity channel [108]. A signal space model is described which represents the average optical power cost and the amplitude constraints of the channel geometrically. In this manner, all time-disjoint signalling schemes can be considered in a common framework. The set of signals satisfying the non-negativity constraint are shown to form a generalized N-cone termed the admissible region. In an analogous fashion, a peak optical power bounding region is defined represents the set of signal vectors satisfying a peak optical power constraint. The geometry of the admissible region is then linked to the peak optical power of each transmittable signal point. The chapter concludes with several examples of the signal space for various choices of basis functions.

4.1 Signal Space of Optical Intensity Signals

As discussed in Section 3.1.1, the optical intensity channel can in some cases be modelled as a baseband electrical channel subject to input amplitude constraints. This section presents a signal space model which retains the same structure of electrical signal spaces, since detection is done in electrical domain, but additionally represents the non-negativity and the average optical power cost geometrically. The properties of the signal space are then explored and related to the set of transmittable points and to the peak optical power of signal points.

4.1.1   Signal Space Model

Consider the case of time-disjoint signalling described in Section 3.1.6, where the support set of all transmitted symbols is the interval $[0, T)$ for some $T > 0$. Using the notation of Section 3.1.2, the transmitter sends one of M signals from the set X. A signal space can be defined by specifying a set of N orthonormal functions, Φ, each time limited to $t \in [0, T)$ such that X $\subset$ span(Φ). Each $x_m(t) \in$ X is represented by the vector $x_m = (x_{m,1}, x_{m,2}, \ldots, x_{m,N})$ with respect to the basis set Φ and the signal constellation is defined as $\Omega = \{x_m : m \in M\}$.

The non-negativity constraint in (3.10) implies that the average amplitude value of the signals transmitted must also be non-negative. In order to represent the average optical cost geometrically, set the function

$$\phi_1(t) = \frac{1}{\sqrt{T}} \, \text{rect}\left(\frac{t}{T}\right) \tag{4.1}$$

where

$$\text{rect}(t) = \begin{cases} 1 & : \quad 0 \leq t < 1 \\ 0 & : \quad \text{otherwise,} \end{cases}$$

as a basis function for every time-disjoint intensity modulation scheme, as shown in Figure 4.1. Note that by assigning $\phi_1(t)$, the dimensionality of the signal space is at most one larger than the dimension of Ω. Due to the orthogonality of the other basis functions,

$$\int_0^T \phi_n(t) dt = \begin{cases} \sqrt{T} & : \quad n = 1 \\ 0 & : \quad 1 < n \leq N. \end{cases} \tag{4.2}$$

The $\phi_1(t)$ basis function represents the average amplitude of each symbol and thus represents the average optical power of each symbol. In this manner, the average optical power requirement is represented in a single dimension.

The average optical power of an intensity signalling set can then be computed from (3.11) as

$$P(\Omega, T) = \frac{1}{\sqrt{T}} P^G(\Omega), \tag{4.3}$$

where $P^G(\Omega)$ is defined as

$$P^G(\Omega) = \sum_{m \in M} \Pr(m) x_{m,1}.$$

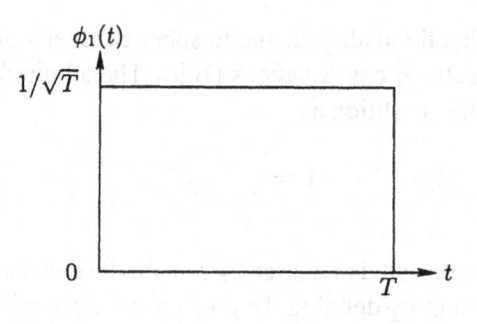

Figure 4.1. The $\phi_1(t)$ basis function.

Thus, P depends both on the symbol interval as well as the expected value of the coordinate value in the ϕ_1 direction. The term P^G can be interpreted as the component of P which depends solely on the constellation geometry.

Unlike the case of the electrical channel where the energy cost of a scheme is completely contained in the geometry of the constellation, the average optical power of an intensity signalling scheme depends on the symbol period. This is due to the fact that $\phi_1(t)$ defined in (4.1) is set to have unit *electrical* energy because detection is done in the electrical domain. The cost constraint for this channel, however, is on the average amplitude. As a result, the average amplitude and hence average optical power depend on T.

4.1.2   Admissible Region

The non-negativity constraint (3.10) is not satisfied by all linear combinations of the elements of Φ. The *admissible region*, Υ, of an optical intensity modulation scheme is defined here as the set of all points satisfying the non-negativity criterion. In terms of the signal space,

$$\Upsilon = \left\{ v \in \mathbb{R}^N : \text{Min}(v) \geq 0 \right\}, \tag{4.4}$$

where for $v = (v_1, v_2, \ldots, v_N)$, $\text{Min} : \mathbb{R}^N \to \mathbb{R}$ is defined as,

$$\text{Min}(v) = \min_{t \in [0,T)} \sum_{n \in N} v_n \phi_n(t).$$

The set Υ is closed, contains the origin and is convex. Convexity can be justified since for any $b_1, b_2 \in \Upsilon$ and any $\alpha \in [0, 1]$, $\alpha b_1 + (1 - \alpha) b_2 \in \Upsilon$ since it describes a non-negative signal.

It is useful to consider cross-sections of Υ for a given ϕ_1 value or equivalently in terms of points of equal average optical power. For any fixed $r \in \mathbb{R}$, $r \geq 0$, define the set,

$$\Upsilon_r = \left\{ (v_1, v_2, \ldots, v_N) \in \Upsilon : v_1 = r \right\} \tag{4.5}$$

as the set of all signal points with a fixed average optical power of $r/\sqrt{T}$. Each of the cross-sections Υ_r forms an equivalence class or *shell* of transmittable symbols. This is directly analogous the to spherical shells of equal energy in the conventional electrical constellations [109]. The admissible region can be written in terms of this partition as

$$\Upsilon = \bigcup_{r \geq 0} \Upsilon_r. \tag{4.6}$$

The set of signals represented in the set Υ_r can also be considered without their common ϕ_1 component by defining the *projection* map $\mathrm{Proj} : \mathbb{R}^N \to \mathbb{R}^N$ that maps $(x_1, x_2, \ldots, x_N)$ to $(0, x_2, x_3, \ldots, x_N)$. Theorem 4.1 summarizes the key properties of Υ.

THEOREM 4.1 *Let* Υ *denote the admissible region of points defined in (4.4).*

1 For $u, v > 0$, $\Upsilon_u = (u/v)\Upsilon_v$.

2 $\Upsilon = \bigcup_{r \geq 0}(r\Upsilon_1)$.

3 Υ_1 *is closed, convex and bounded.*

4 If $\partial\Upsilon_1$ *denotes the set of boundary points of* Υ_1, *then*

   (i) $0 \notin \mathrm{Proj}(\partial\Upsilon_1)$ *and*

   (ii) $\partial\Upsilon_1 = \{v \in \Upsilon_1 : \mathrm{Min}(v) = 0\}$.

5 Υ *is the convex hull of a generalized N-cone with vertex at the origin, opening about the* ϕ_1-*axis and limited to* $\phi_1 \geq 0$.

Proof : (the details can be omitted without loss of continuity)
Property 1. The collection of signals $(v/u)\Upsilon_u$ is a set of pulses with average optical power $v/\sqrt{T}$. So, $(v/u)\Upsilon_u \subseteq \Upsilon_v$. Similarly, $(u/v)\Upsilon_v \subseteq \Upsilon_u$, which implies that $\Upsilon_v \subseteq (v/u)\Upsilon_u$.
Property 2. Follows directly from (4.6) and Property 1.
Property 3. The set Υ_1 is closed by (4.4). For $v_1, v_2 \in \Upsilon_1$ and $\alpha \in [0, 1]$ the average optical amplitude value of $x = \alpha v_1 + (1 - \alpha)v_2$ is $1/\sqrt{T}$. Hence, $x \in \Upsilon_1$ implies Υ_1 is convex.
A set in $\mathbb{R}^N$ is *bounded* if it is contained in an N-ball of finite radius. The region $\mathrm{Proj}(\Upsilon_1)$ is

$$\mathrm{Proj}(\Upsilon_1) = \left\{ v = (v_1, v_2, \ldots, v_N) \in \mathbb{R}^N : v_1 = 0, \ \mathrm{Min}(v) \geq -\frac{1}{\sqrt{T}} \right\}.$$

The signals in $\mathrm{Proj}(\Upsilon_1)$ have zero average amplitude in $[0, T)$, by definition (4.2). The set $\mathrm{Proj}(\Upsilon_1)$ is closed and contains the origin. Consider $v \in \mathrm{Proj}(\Upsilon_1)$ such that $\|v\| = q$ for some $q > 0$. If there is no such point, then $\mathrm{Proj}(\Upsilon_1)$ is contained within ball of radius q since Υ_1 convex. Otherwise,

$k\upsilon \in \text{Proj}(\Upsilon_1)$ for $k \in [0, |1/(\sqrt{T}\text{Min}(\upsilon))|]$ is contained in an N-ball of radius greater than $|q/(\sqrt{T}\text{Min}(\upsilon))|$. The union of all such N-balls for all $\upsilon \in \text{Proj}(\Upsilon_1)$ contains $\text{Proj}(\Upsilon_1)$ implying that $\text{Proj}(\Upsilon_1)$ is bounded. Since the ϕ_1 coordinate of all points in Υ_1 is the same, $\text{Proj}(\Upsilon_1)$ bounded implies Υ_1 is bounded.

Property 4. Denote the set of boundary points of Υ_1 as $\partial\Upsilon_1$. An $(N-1)$-ball of radius $\epsilon > 0$ in $\text{Proj}(\Upsilon_1)$ exists about the origin since otherwise it would imply that a signal point $x \in \text{Proj}(\Upsilon_1)$, $x \neq 0$, were either non-negative or non-positive, which is impossible due to the construction of the signal space in (4.2). Therefore, $0 \notin \text{Proj}(\partial\Upsilon_1)$.

From Property 3, for $\upsilon \in \text{Proj}(\Upsilon_1)$, $\upsilon \neq 0$, $k\upsilon \in \text{Proj}(\Upsilon_1)$ for $k \in [0, |1/(\sqrt{T}\text{Min}(\upsilon))|]$. The boundary points are those for which k is maximized. The set $\text{Proj}(\partial\Upsilon_1)$ is the set of extremal points with minimum amplitude equal to $-1/\sqrt{T}$. Using the inverse map, $\partial\Upsilon_1$ is the set of points in Υ_1 with minimum amplitude equal to zero.

Property 5. A generalized N-cone is defined as a surface in $\mathbb{R}^N$ which is parameterized as $C(u, v) = c + vC'(u)$, where c is a fixed vector called the vertex of the cone and $C'(u)$ is a curve in $\mathbb{R}^N$ and u and v are parameters [110]. Using Properties 1 and 2, $\partial\Upsilon$ can be written as $\partial\Upsilon_r = r\partial\Upsilon_1$, for $r \geq 0$. The set of boundary points $\partial\Upsilon$ satisfies the definition and is a generalized cone with vertex at the origin opening about the ϕ_1-axis. Since Υ_r is the convex hull of $\partial\Upsilon_r$, Υ is the convex hull of a generalized cone limited to the half-space $\phi_1 \geq 0$. □

Figure 4.2 presents a plot of a portion of the admissible region for 3-dimensional raised-QAM scheme defined in Section 4.2. Notice that the region forms a 3-dimensional cone with disc shaped cross-sections in ϕ_1, satisfying the properties of Theorem 4.1.

4.1.3   Peak Optical Power Bounding Region

In wireless optical channels, the average optical power constraint typically dominates the peak power constraint, as discuss in Section 2.1.2. However, in any practical system the peak optical power must also be limited.

The *peak optical power bounding region*, $\Pi(p)$, is defined as the set of points corresponding to signals that have amplitudes bounded above by $p/\sqrt{T}$. That is, for some $p \geq 0$,

$$\Pi(p) = \left\{ \pi \in \mathbb{R}^N : \text{Max}(\pi) \leq \frac{p}{\sqrt{T}} \right\}, \tag{4.7}$$

where for $\pi = (\pi_1, \pi_2, \ldots, \pi_N)$, $\text{Max} : \mathbb{R}^N \to \mathbb{R}$ is defined as,

$$\text{Max}(\pi) = \max_{t \in [0,T)} \sum_{n \in \mathbb{N}} \pi_n \phi_n(t).$$

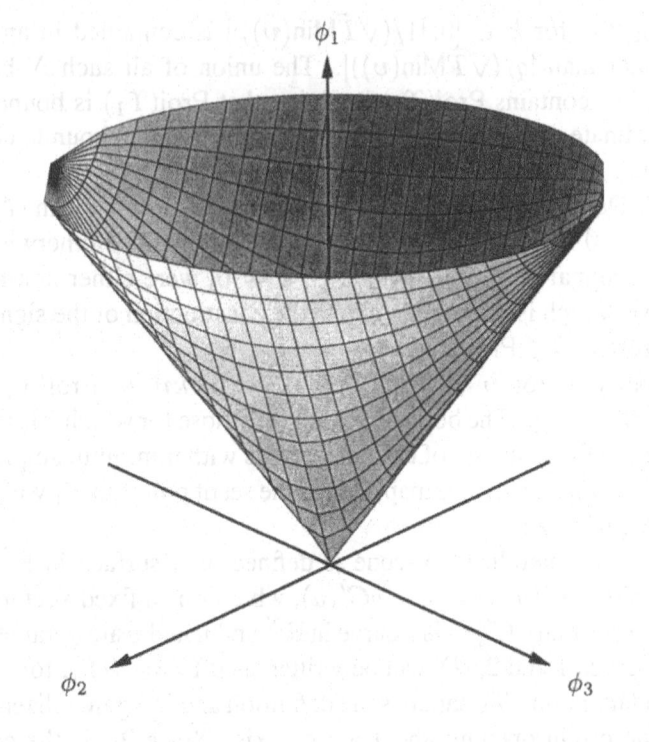

Figure 4.2. Three-dimensional admissible region, Υ, for the raised-QAM example in Section 4.2.

Interestingly, Υ and $\Pi(p)$ are closely related and their explicit relation is illustrated in the following theorem.

THEOREM 4.2 *The peak optical power bounding region* $\Pi(p) = -\Upsilon + p\boldsymbol{\phi}_1$, *where* $\boldsymbol{\phi}_1$ *is a unit vector in the* ϕ_1 *direction.*

Proof : By definition (4.4), Υ is the set of signals with non-negative amplitudes. The set $-\Upsilon$ is the set of signals for which the maximum possible amplitude is zero. Since $\phi_1(t)$ is constant in a symbol period, the addition of $p\phi_1(t), p \geq 0$, to each signal in $-\Upsilon$ yields the set of signals with maximum at most $p/\sqrt{T}$. The region $\Pi(p)$ is then given as

$$\Pi(p) = -\Upsilon + p\boldsymbol{\phi}_1. \tag{4.8}$$

$\square$

Since $\Pi(p)$ differs from Υ by an affine transform, it is clear that $\Pi(p)$ is also the convex hull of an N-dimensional generalized cone with vertex at $(p, 0, 0, \ldots, 0)$ and opening about the negative ϕ_1-axis. Figure 4.3 illustrates a portion of the peak bounding region for the 3-D raised-QAM example in Section 4.2.

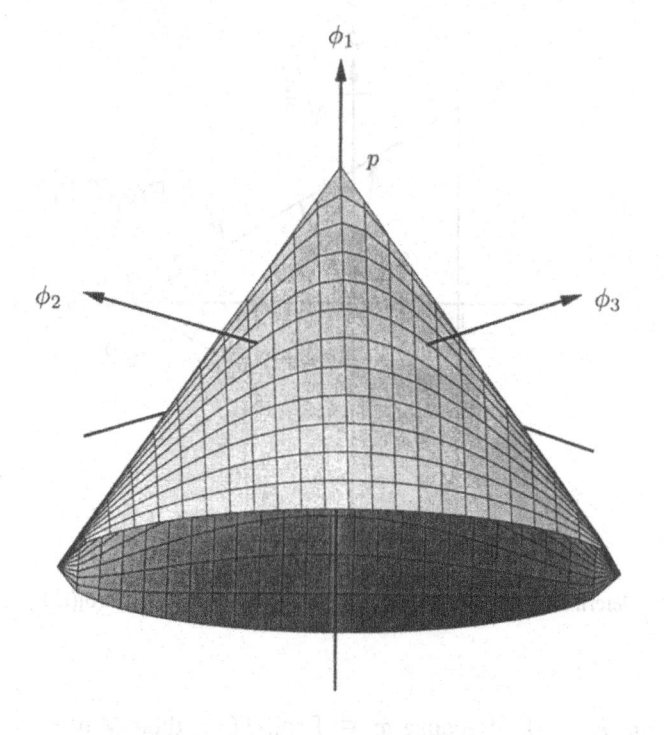

Figure 4.3. Three dimensional peak bounding region, $\Pi(p)$, for the raised-QAM example in Section 4.2.

4.1.4   Peak Optical Power per Symbol

Although we are able to specify the region of points which satisfy the peak constraint, $\Pi(p)$, this set does not reveal any details about the peak amplitude of the signal in the set. In the construction of modulation schemes, it would be useful to have some knowledge of which points have high optical peak values.

The region Υ can be completely characterized by looking at a single cross-section as shown in Property 2 of Theorem 4.1. Therefore, finding the peak values of signals in Υ_1 will give the peak values of all points in Υ_r through scaling. As discussed earlier, the signals with the maximum peak amplitudes in Υ_1 must lie in $\partial\Upsilon_1$.

For $x \in \text{Proj}(\partial\Upsilon_1)$, $\text{Min}(x) = -1/\sqrt{T}$ by Theorem 4.1. The signal $-x$ has $\text{Min}(-x) = \text{Max}(x)$ and $\text{Max}(-x) = 1/\sqrt{T}$. Define the vector $\hat{x} \in \text{Proj}(\partial\Upsilon_1)$ such that

$$\hat{x} = -kx, \tag{4.9}$$

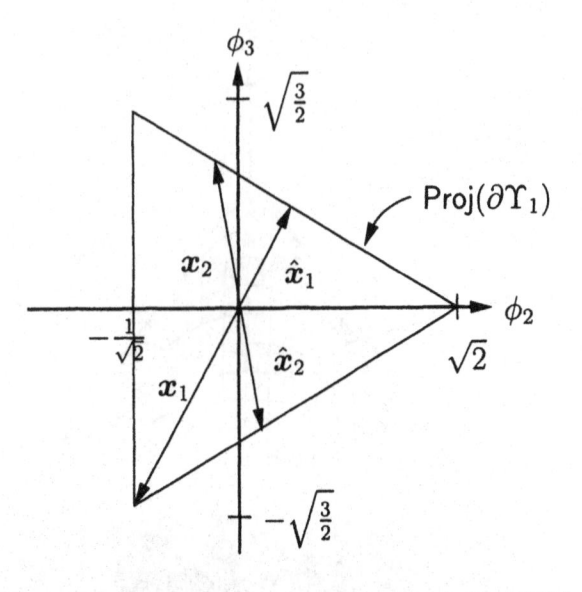

Figure 4.4. Determining the peak amplitude values of elements in $\mathsf{Proj}(\partial \Upsilon_1)$ for 3-D PAM bases.

for a unique $k > 0$. Because $\hat{x} \in \mathsf{Proj}(\partial \Upsilon_1)$, then $\mathsf{Min}(\hat{x}) = -1/\sqrt{T}$, however, $\mathsf{Min}(\hat{x}) = -k\mathsf{Max}(x)$. Therefore,

$$\mathsf{Max}(x) = \frac{k^{-1}}{\sqrt{T}}. \tag{4.10}$$

Using (4.9) the above simplifies to

$$\mathsf{Max}(x) = \frac{\|x\|}{\|\hat{x}\|} \cdot \frac{1}{\sqrt{T}}, \tag{4.11}$$

which exists since $\|\hat{x}\| > 0$. Equation (4.10) implies that $\mathsf{Max}(\hat{x}) = k/\sqrt{T}$. Figure 4.4 illustrates the scenario in (4.11) for the 3-D PAM basis defined in Section 4.2. From the figure, it is possible to deduce that x_1 has a larger peak value than x_2 and that $\mathsf{Max}(x_2) = \mathsf{Max}(\hat{x}_2)$ by observing the relative magnitudes of the vectors. Since $\mathsf{Proj}(\Upsilon_1)$ is convex, the peak optical power values for all signals in $\mathsf{Proj}(\Upsilon_1)$ differ from those in (4.11) by a scaling factor in the interval $[0, 1]$.

The peak optical power of the points in $v \in \Upsilon_1$ is

$$\mathsf{Max}(v) = \mathsf{Max}(\mathsf{Proj}(v)) + \frac{1}{\sqrt{T}}. \tag{4.12}$$

The *peak-to-average* optical power ratio, (PAR), for all $v \in \Upsilon_1$ can be computed using (4.11) and (4.12) as

$$\text{PAR}(\Upsilon_1) = \max_{x \in \text{Proj}(\partial \Upsilon_1)} \frac{\|x\|}{\|\hat{x}\|} + 1, \tag{4.13}$$

with $\hat{x}$ as defined in (4.9). Maximization is done over points in $\text{Proj}(\partial \Upsilon_1)$ since the signal with the largest peak value must be in $\text{Proj}(\partial \Upsilon_1)$.

The peak optical power related to the signal vectors in Υ is found by scaling the peak values found for $v \in \Upsilon_1$. The PAR of points in Υ_r will be the same as (4.13) since the average and peak optical power scale by the factor r.

4.1.5 Peak-Symmetric Signalling Schemes

A basis set Φ is termed *peak-symmetric* if $\text{Proj}(\Upsilon_1)$ is closed under inversion. From the point of view of signal amplitudes, using (4.9) and (4.11), this condition implies that $x = -\hat{x}$ and $\text{Max}(x) = \text{Max}(-x)$ for $x, \hat{x} \in \text{Proj}(\partial \Upsilon_1)$ and hence the term peak-symmetric. Furthermore, for a peak-symmetric scheme, using (4.12), for $v \in \Upsilon_1$

$$\text{Max}(v) = \frac{2}{\sqrt{T}}. \tag{4.14}$$

As discussed in Section 4.1.4, this maximum amplitude is achieved for $v \in \partial \Upsilon_1$. Note that the 3-D PAM scheme in Figure 4.4 is not peak-symmetric. Section 4.2.2 presents examples of the $\text{Proj}(\Upsilon_1)$ of 3-D peak-symmetric schemes.

Peak-symmetric schemes are desirable in the sense that the maximum amplitude value in Υ_1 is achieved by all points in $\partial \Upsilon_1$. Maximization over all points in $\text{Proj}(\partial \Upsilon_1)$ is not required in the calculation of the PAR (4.13) since it is satisfied at every point. Thus, peak-symmetric signalling schemes fully exploit a given peak constraint by maximizing the number of points achieving the peak limit.

4.2 Examples

4.2.1 Definition of Example Schemes

As noted in Section 4.1.1 and in [88], the performance of optical intensity modulation techniques depends not only on the electrical energy of the pulses (i.e., the geometry of the signal space), but also on the pulse shapes chosen to define the space. This section defines the basis functions used to form signals in the example schemes considered. Note that all symbols are limited to the interval $t \in [0, T)$ and $\phi_1(t)$ is specified as in (4.1).

Quadrature pulse amplitude modulation (QAM) is a familiar modulation scheme in wireless communications. Here we define the optical intensity modulation scheme *raised-QAM* as a 3-dimensional modulation scheme with basis

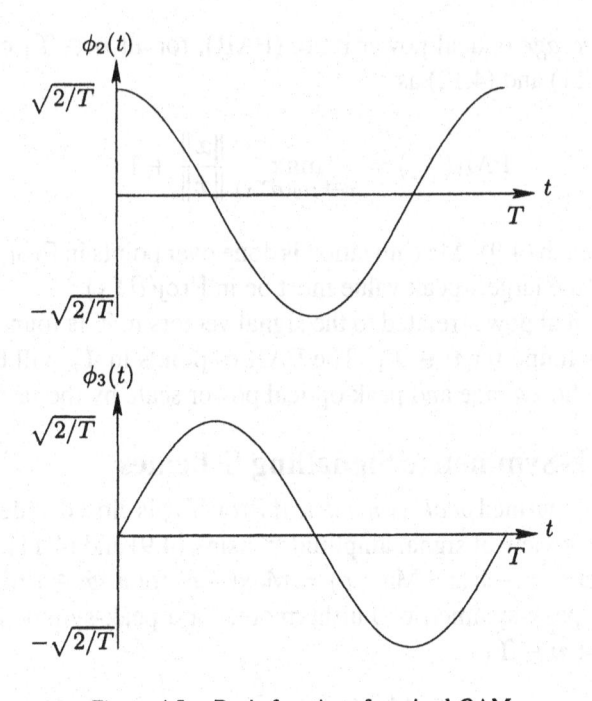

Figure 4.5. Basis functions for raised-QAM.

functions

$$\phi_2(t) = \sqrt{\frac{2}{T}} \cos(2\pi t/T)$$

$$\phi_3(t) = \sqrt{\frac{2}{T}} \sin(2\pi t/T).$$

Figure 4.5 plots the basis function for raised-QAM. These bases are similar to the QAM bases presented in Section 3.3.2 with $q = 1$, however, in raised-QAM there is no assumption that the average amplitude is fixed over all transmitted symbols.

Adaptively-biased QAM (AB-QAM) [104, 5] is a three dimensional modulation scheme which is defined using the basis functions

$$\phi_2(t) = \frac{1}{\sqrt{T}} \text{rect}(t/T) - \frac{2}{\sqrt{T}} \text{rect}(2t/T - 1/2)$$

$$\phi_3(t) = \frac{1}{\sqrt{T}} \text{rect}(t/T) - \frac{2}{\sqrt{T}} \text{rect}(2t/T - 1).$$

More generally, these functions are *Walsh functions*. This characterization is especially useful in light of the signal space definition in Section 4.1.1, since the basis functions of AB-QAM are scaled and shifted versions of the first three Walsh functions [111]. Figure 4.6(a) plots the basis functions for AB-QAM.

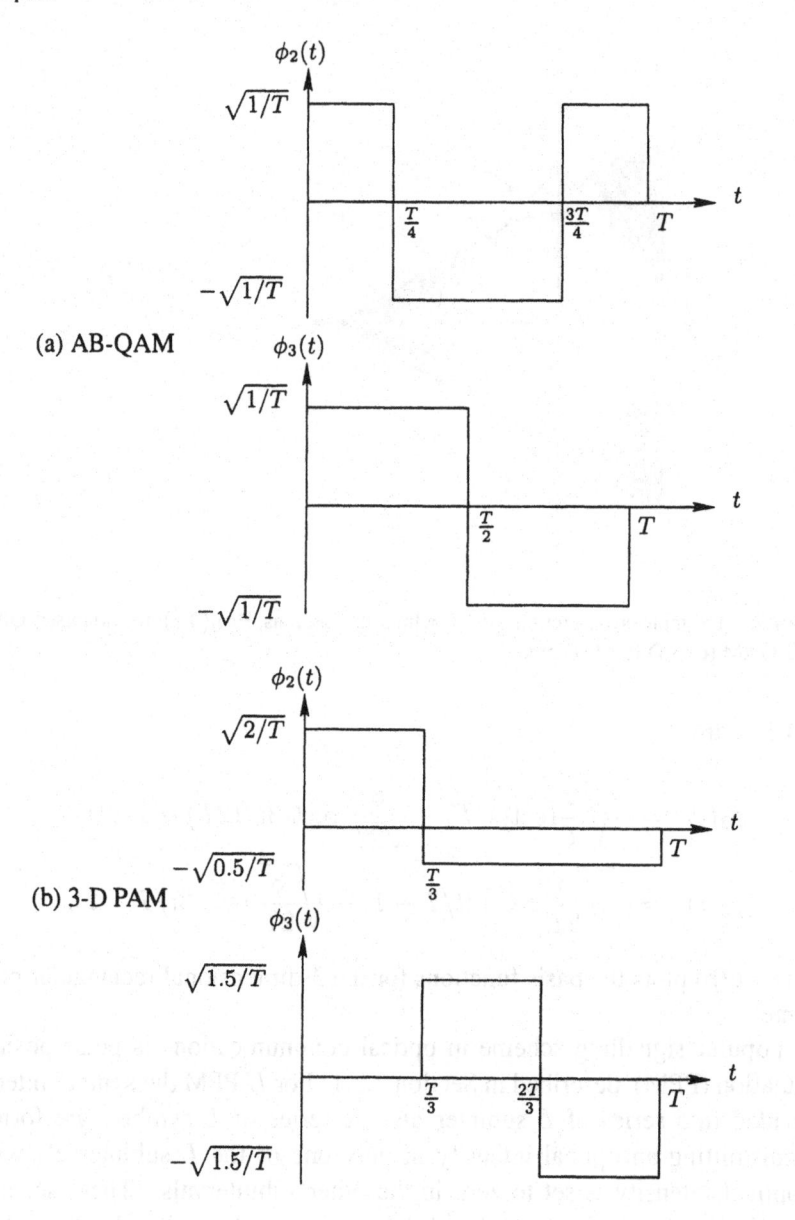

Figure 4.6.   Basis functions for (a) AB-QAM and (b) 3-dimensional rectangular pulse scheme.

A 3-D PAM scheme can be constructed by transmitting three one-dimensional rectangular pulse shaped PAM symbols. This construction is analogous to the techniques used in conventional lattice coding literature [112] and is the case considered in earlier optical lattice coding work [79]. The basis functions for this 3-dimensional scheme, according to the signal space model defined in Sec-

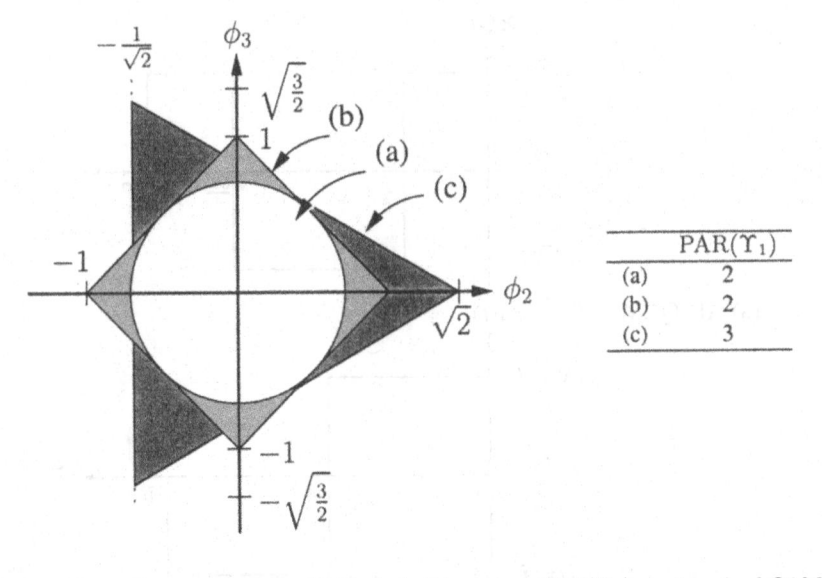

	PAR(Υ_1)
(a)	2
(b)	2
(c)	3

Figure 4.7. Example cross-sections of 3-D admissible regions, Proj(Υ_1), for (a) raised-QAM, (b) AB-QAM (c) 3-D PAM scheme.

tion 4.1.1, are

$$\phi_2(t) = \sqrt{\frac{2}{T}}\mathrm{rect}(t/T) - \frac{3}{\sqrt{2T}}\,\mathrm{rect}(3t/(2T) - 1/2))$$

$$\phi_3(t) = \sqrt{\frac{3}{2T}}\,\mathrm{rect}(3t/T - 1) - \sqrt{\frac{3}{2T}}\,\mathrm{rect}(3t/T - 2).$$

Figure 4.6(b) plots the basis functions for the 3-dimensional rectangular pulse scheme.

A popular signalling scheme in optical communications is pulse position modulation (PPM), described in Section 3.3.1. For L-PPM the symbol interval is divided into series of L subintervals. A series of L symbols are formed by transmitting an optical intensity in only one of the L subintervals while the optical intensity is set to zero in the other subintervals. These schemes were originally conceived for the photon-counting channel and achieve high power efficiency at the expense of bandwidth efficiency [113]. Note that a PPM modulation scheme can be thought of as a coded version of the 3-D PAM scheme discussed earlier.

4.2.2 Geometric Properties

Figure 4.7 contains plots of the regions Proj(Υ_1) of the example bases defined in Section 4.2.1. As predicted by Property 3 of Theorem 4.1 the regions are all closed, convex and bounded.

In the case of raised-QAM, every point in $\text{Proj}(\Upsilon_1)$ represents a sinusoid time limited to $[0, T)$ with amplitude determined by the square distance from the origin. Since sinusoids of the same energy have the same amplitude, regardless of phase, all points equidistant from the origin have the same amplitude. As a result, a 2-D disc naturally results as the region $\text{Proj}(\Upsilon_1)$. In the case of 3-D PAM, $\text{Proj}(\Upsilon_1)$ is an equilateral triangle with sides of length $\sqrt{6}$. It is easy to show that the signals corresponding to a 3-PPM scheme are represented by the vertices of the triangle. In this manner, PPM can be seen as a special case of the 3-D PAM scheme.

Note that the raised-QAM and AB-QAM scheme, (a) and (b) in Figure 4.7, both represent peak-symmetric signalling schemes. While the $\text{Proj}(\Upsilon_1)$ regions for these two modulation schemes are different, the peak-to-average amplitude value of the signals represented is 2 in both cases as given in (4.14). The 3-D PAM scheme is not peak-symmetric. The largest peak values occur for the points at the vertices of the triangle (x_1 in Figure 4.4) to give a PAR of 3 for Υ_1.

4.3 Conclusions

In this chapter a signal space model is defined for time-disjoint signalling schemes on the wireless optical intensity channel. This channel model represents the non-negativity constraint geometrically as an admissible region of points. This set is shown to be the convex hull of a generalized cone. Similarly, the set of points satisfying a peak optical power constraint are also shown to reside within a similar generalized cone. A key feature of the model is that it represents the average optical power of each symbol as the coordinate value in a given direction. This feature allows for the construction of modulation subject to an optical power limitation to be done based on the geometry of the signal space. Some example basis sets are defined and the associated regions presented.

In Chapters 5 and 6 this signal space model will be used to define modulation schemes as well as to compute bounds on the capacity for a given basis set.

Chapter 5

LATTICE CODES

The construction of multi-level, multi-dimensional constellations for bandwidth limited electrical channels using the formalism of lattice codes has been explored extensively in the literature [114, 115, 112, 116, 117, 109, 118, 119]. However, the study of lattice codes on optical intensity channels has received considerably less attention.

Fiber optical channels are typically considered as being power-limited rather than bandwidth-limited. The case of signalling over bandwidth limited optical channels, such as some wireless optical links, has not received much attention. Pioneering work in signalling design for the optical intensity channels noted that unlike the electrical channels, where the energy of the signals is important, the shape of the pulses used in transmission as well as the energy determine the performance of optical intensity schemes [88]. Shiu and Kahn developed the initial work on lattice codes for wireless optical intensity channels by constructing higher dimensional modulation schemes from a series of one-dimensional constituent constellations of rectangular on-off keying [79]. However, they used the first-null measure of bandwidth and the effects of constellation shaping on this bandwidth were not considered.

This chapter uses the signal space model of Chapter 4 to construct lattice codes satisfying the amplitude constraints of the optical intensity channel [108]. An optical constellation figure of merit is defined and optical power gain with respect to a baseline is computed. Optimal shaping regions in the sense of minimum average optical power are derived and optical power gain is computed. The impact of a peak optical power constraint is investigated and gains computed. This chapter concludes with a comparison of designed lattice codes on an idealized point-to-point link. Spectrally efficient modulation is shown to outperform rectangular PPM for short range links.

5.1 Definition of Lattice Codes

5.1.1 Background

As discussed in Chapter 3, for high SNR channels, the minimum Euclidean distance, $d_{\min}$, is an important metric in determining the error performance of a signalling scheme. Also, recall that when an orthonormal basis is selected for the signal space that the squared-norm of any signal vector is the energy of the signal. In electrical channels, the goal of constellation design is, for a given basis set, to find a constellation which for a given $d_{\min}$ minimizes the average energy of the modulation scheme. This problem is equivalent to finding an arrangement or *packing* for a set of N-dimensional spheres of radius $d_{\min}$ so that they are packed as tightly as possible around the origin without intersecting. To the knowledge of the author, the densest or most efficient packings have only been found for spaces of dimension at most 3 and the problem of finding the densest packing in higher dimensions remains an open problem [120].

A *lattice*, Λ, is type of packing which includes the origin and if there are points $u, v \in \Lambda$ then $u + v \in \Lambda$ and $u - v \in \Lambda$. In other words, Λ is an additive *group* which is also a sub-group of $\mathbb{R}^n$. Figure 5.1 illustrates the integer, Z_1, and the hexagonal, A_2, lattice packings. Every lattice has a *fundamental region*. The fundamental region is such that if it is repeated such that a single lattice point is in each region, it complete covers, or *tessellates*, the space. The hexagon in Figure 5.1 is a fundamental region for A_2. Note that the fundamental region is not unique, however, the volume of any fundamental region, $V(\Lambda)$, is the same regardless of how the fundamental region is chosen. Some other common measures of a lattice are $d_{\min}^2$, the minimum square Euclidean distance between any two lattice points, and $K_{\min}$, the "kissing-number" or the number of nearest neighbors to any given lattice point. A normalized lattice packing density known as *Hermite's parameter*, $\gamma_n(\Lambda)$ [120], can also be defined and takes the form,

$$\gamma_n(\Lambda) = \frac{d_{\min}^2}{V(\Lambda)^{2/N}}. \tag{5.1}$$

This satisfies intuition, since for a given $d_{\min}$, dense lattices minimize the volume of the fundamental region assigned to each lattice point. The densest lattices in one and two dimensions are Z_1 and A_2 respectively.

An N-dimensional *lattice code* is constructed by intersecting a lattice or in general a translate of a lattice, $\Lambda + t$, with a shaping region, Θ, which selects a finite number of points, i.e.,

$$\Omega = (\Lambda + t) \cap \Theta,$$

where t is some vector. The lattice is chosen to provide a set of points with large minimum distance for a given fundamental volume, i.e., large γ_n, whereas the shaping region is designed so as to minimize to cost of the subset of points

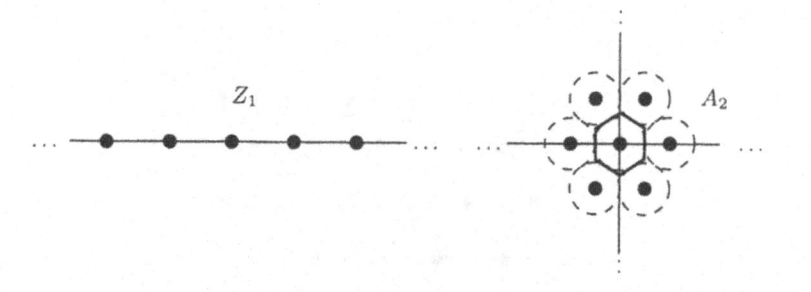

Figure 5.1. Example one-dimensional integer lattice and two-dimensional hexagonal lattice.

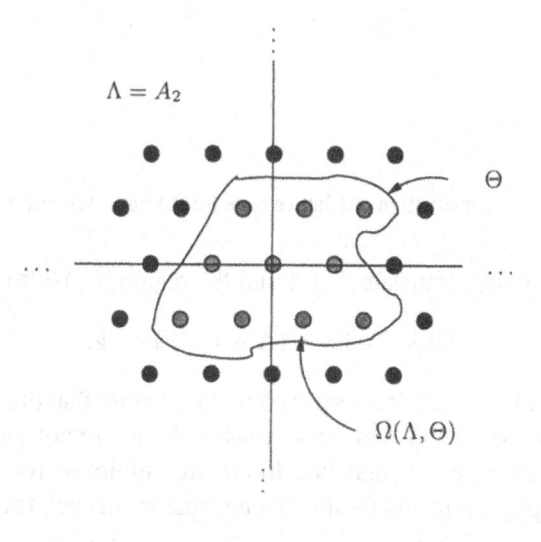

Figure 5.2. Construction of a lattice code.

chosen [109]. Figure 5.2 illustrates the construction of a lattice code in which points inside the shaping region are selected as the constellation. The optimum shaping region, in the sense of minimizing the average energy of the constellation is an N-sphere. However, as N grows large, the size of the two-dimensional sub-constellations used to construct the N dimensional constellation grows as N. The optimum trade-off between this constellation expansion and shaping gain results in a truncated polydisc shaping region [118].

5.1.2 Lattice Codes for Optical Intensity Channels

The definition of lattice codes in the electrical case does not take into account the amplitude constraints of the optical intensity channel. Lattice codes satisfying the constraints of the optical intensity channel, however, can be defined for a given basis set Φ using the signal space model in Section 4.1.1.

The *shaping region*, Ψ, is defined so that $\Theta = \Upsilon \cap \Psi$ is bounded. An N-dimensional lattice constellation is formed through the intersection of an

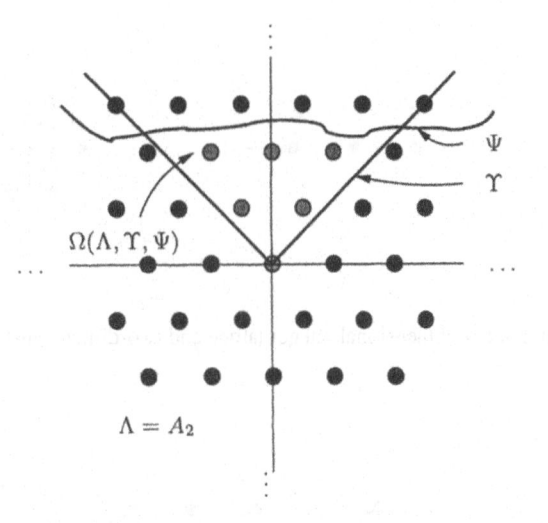

Figure 5.3. Construction of a lattice code for the optical intensity channel.

N-dimensional lattice translate $\Lambda + t$ and the region $\Upsilon \cap \Psi$ to give,

$$\Omega(\Lambda, \Upsilon, \Psi) = (\Lambda + t) \cap \Upsilon \cap \Psi, \qquad (5.2)$$

as illustrated in Figure 5.3. It is assumed in this chapter that the symbols $x \in \Omega$ are selected independently and equiprobably from symbol period to symbol period. At the receiver, N matched filters are employed for detection. The channel decomposes into an N-dimensional vector channel, following (3.5), as

$$y = x + n$$

where $x \in \Omega$ and n is Gaussian distributed with independent components and variance σ^2 per dimension. As discussed in Section 3.1.2, the pair (Ω, Φ) is used to denote a modulation scheme since it describes the set X of transmittable signals.

In the following sections the performance of optical lattice codes is related to selection of the sets of Λ, Υ and Ψ.

5.2   Constellation Figure of Merit, Gain

The *constellation figure of merit* (CFM) is a popular measure of the energy efficiency of a signalling scheme in electrical channels and is defined as

$$\mathrm{CFM}(\Omega) = \frac{d_{\min}^2(\Omega)}{E(\Omega)}$$

where $d_{\min}(\Omega)$ is the minimum Euclidean distance between constellation points and $E(\Omega)$ is average energy of the constellation [112]. It is desirable for an Ω

to possess a large CFM, that is, for a given d_{min} to have a small average energy cost. It is important to note that the CFM for electrical channels depends solely on the *geometry* of the constellation.

An analogous measure for optical intensity channels which quantifies the *optical* power efficiency of the scheme is [5, 79]

$$\text{CFM}(\Omega, T) = \frac{d_{min}(\Omega)}{P(\Omega, T)}$$

$$= \sqrt{T} \cdot \frac{d_{min}(\Omega)}{P^G(\Omega)}, \tag{5.3}$$

where $P(\Omega, T)$ is average optical power (4.3). The CFM in (5.3) is invariant to scaling of the constellation as in the case in electrical channels [112]. The optical CFM is unaffected by L-fold Cartesian product of Ω so long as the symbol period also increases L-fold, i.e., $\text{CFM}(\Omega^L, LT) = \text{CFM}(\Omega, T)$. The CFM in electrical channels is invariant under orthogonal transformations of Ω [121], whereas $\text{CFM}(\Omega, T)$ is invariant under a subset of orthogonal transformations which leave the ϕ_1 coordinate unaffected. Additionally, the CFM in (5.3) depends on T via (4.3) and is not unitless as is the case in conventional electrical channels.

The CFM provides a means to computed the optical power gain of one scheme versus another. The probability of a symbol error, P_e, can be approximated for a given constellation Ω as

$$P_e(P(\Omega, T)/\sigma) \approx \bar{N}(\Omega) \cdot Q\left(\frac{\text{CFM}(\Omega, T) \cdot P(\Omega, T)}{2\sigma}\right)$$

as discussed in (3.7). Let $P_e^{-1}(q; \Omega, T)$ evaluate to the $P(\Omega, T)/\sigma$ required to achieve a symbol error rate of q. The optical power gain, G_q, of (Ω, Φ) with respect to some baseline $(\Omega_\oplus, \Phi_\oplus)$ with symbol period $T_\oplus$ operating over the same channel and at the same probability of symbol error, q, is given by the ratio $P(\Omega_\oplus, T_\oplus)/P(\Omega, T)$ or

$$G_q = P_e^{-1}(q; \Omega_\oplus, T_\oplus)/P_e^{-1}(q; \Omega, T)$$
$$= \text{CFM}(\Omega, T)/\text{CFM}(\Omega_\oplus, T_\oplus) \times [Q^{-1}(q/\bar{N}(\Omega_\oplus)/Q^{-1}(q/\bar{N}(\Omega))].$$

The asymptotic optical power gain of (Ω, Φ) over $(\Omega_\oplus, \Phi_\oplus)$, in the limit as $q \to 0$, can be shown to be

$$G(\Omega, T, T_\oplus) = \lim_{q \to 0} G_p = \text{CFM}(\Omega, T)/\text{CFM}(\Omega_\oplus, T_\oplus),$$

which is independent of the error coefficients and noise variance [109].

5.3   Baseline Constellation

In order to compute a gain, a common baseline must be taken for comparison. An M-PAM constellation, $\Omega_\oplus$, with a rectangular basis function is taken as the baseline. This constellation can be formed following (5.2) where $\Lambda = d_{min} Z_1$, $t = 0$, $\Upsilon = [0, \infty)$ and $\Psi = [0, (M-1)d_{min}]$. The resulting modulation scheme is denoted $(\Omega_\oplus, \Phi_\oplus)$ where the single basis function in $\Phi_\oplus$ is $\phi_1(t)$ (4.1) with symbol period $T_\oplus$. Under the assumption that all constellation points are chosen equiprobably the baseline CFM is,

$$\mathrm{CFM}(\Omega_\oplus, T_\oplus) = \frac{2}{|\Omega_\oplus| - 1}\sqrt{T_\oplus}.$$

The constellation $\Omega_\oplus^L$ is formed through the L-fold Cartesian product of $\Omega_\oplus$ with itself and can be realized by transmitting a series of L symbols of $\Omega_\oplus$ each of fixed symbol period $T_\oplus$. Since $\mathrm{CFM}(\Omega_\oplus^L, LT_\oplus) = \mathrm{CFM}(\Omega_\oplus, T_\oplus)$, and there is no asymptotic optical power gain.

The asymptotic optical power gain of signalling scheme (Ω, Φ) over this baseline can then be computed as

$$
\begin{aligned}
G(\Omega, T, T_\oplus) &= \frac{\mathrm{CFM}(\Omega, T)}{\mathrm{CFM}(\Omega_\oplus, T_\oplus)} \\
&= \frac{|\Omega_\oplus| - 1}{2}\sqrt{\frac{T}{T_\oplus}\frac{d_{min}(\Omega)}{P^G(\Omega)}}.
\end{aligned}
\tag{5.4}
$$

5.4   Spectral Considerations

In an comparison of signalling schemes, the spectral properties of each must be considered. Two schemes are compared on the basis of having equal *bandwidth efficiencies*,

$$\eta = \frac{R}{W}, \tag{5.5}$$

where $R = \log_2(M)/T$ is the bit rate of the data source in bits per second and W is a measure of the bandwidth support required by the scheme, as discussed in Section 3.2.

In the case of lattice codes for electrical channels, it is assumed that the N-dimensional constellations under consideration are formed by a series of $N/2$ constituent 2-dimensional QAM symbols. Furthermore, each constituent QAM symbol is assumed to be shaped using ideal Nyquist pulse shaping, which yields a strictly bandlimited power spectral density with bandwidth $W = 1/T$ Hz, for a symbol interval of $(N/2)T$. Thus, assuming equiprobable signalling, the bandwidth efficiency of (Ω, Φ) reduces to

$$\eta = \frac{2}{N}\log_2 |\Omega|. \tag{5.6}$$

This is often referred to as the *normalized bit rate* per two dimensions [112, 109]. To compare two schemes, it is necessary to equate their normalized bit rates. It should be noted that the spectral characteristics of (Ω, Φ) have been completely represented in terms of the geometry of the constellation, and are independent of the signalling interval or bandwidth. The impact of constellation shaping on the power spectral density of a scheme, at a given rate, is typically not considered. Indeed, in the case of block-coded modulation, using constituent 2-D QAM symbols, sufficient conditions on the constituent codes for no spectral shaping have been derived [122].

In general, the power spectral density (3.12), depends on two factors : the pulse shapes, through the $x_i^F(f)$, and on the correlation between symbols. In the case of bandlimited $x_i^F(f)$, the resulting $S_X^c(f)$ must necessarily be bandlimited since it arises due to the sum of a set of similarly bandlimited functions in (3.12). Thus, if bandlimited basis functions are used, the resulting power spectral density will be bandlimited independent of the shaping or coding.

As discussed in Section 3.1.6, the definition of the bandwidth for time-disjoint signalling is not obvious. In previous work on modulation schemes and lattice codes for optical intensity channels, the first spectral null of the power spectral density was used as a bandwidth measure [79, 11, 77]. In this chapter the fractional power bandwidth (3.15) is used to define the bandwidth of signalling schemes. This bandwidth measure is chosen since it better captures the impact of coding and shaping on the spectral characteristics of the channel.

To compare optical intensity schemes (Ω, Φ) versus $(\Omega_\oplus, \Phi_\oplus)$, the bandwidth efficiencies must be equal.

Since the average optical power depends on T, as shown in (4.3), changes in T will change the average optical power, P, but leave the constellation geometry unaffected because the basis functions are scaled to have unit electrical energy independent of T. Thus, unlike electrical channels, the geometry of the constellation does *not* completely represent the average optical power of the scheme. It is therefore necessary to fix both the bandwidth and the bit rate of (Ω, Φ) and $(\Omega_\oplus, \Phi_\oplus)$ to be equal in order to have a fair comparison.

Fixing the rate of the schemes,

$$\log_2 |\Omega_\oplus| = \frac{T_\oplus}{T} \log_2 |\Omega|. \tag{5.7}$$

Define $\kappa = 2W_K T$ for (Ω, Φ) and define $\kappa_\oplus = 2W_{K\oplus}T_\oplus$, where $W_{K\oplus}$ is the fractional power bandwidth of the baseline scheme. Writing W_K and $W_{K\oplus}$ in terms of κ and $\kappa_\oplus$ and equating them gives the definition

$$\nu = \frac{\kappa}{\kappa_\oplus} = \frac{T}{T_\oplus}. \tag{5.8}$$

Combining the results of (5.7) and (5.8) gives,

$$\log_2 |\Omega_\oplus| = \frac{1}{\nu} \log_2 |\Omega|. \tag{5.9}$$

The term κ can be interpreted as the "essential" dimension of (Ω, Φ) where signals are time-limited to $[0, T)$ and the fractional power bandwidth is W_K. This dimension is analogous to the Landau-Pollak dimension, defined in Section 3.2, for time-limited signals. Although this analogy is not rigorous in the case of power spectra, nonetheless, it provides interesting insight into the resulting expressions.

If κ is interpreted as the effective dimension of Ω, the parameter ν in (5.8) is then the effective number of dimensions of Ω with respect to the baseline. Equation (5.9) can then be interpreted as the *effective normalized rate* in units of bits per effective baseline dimension. This interpretation is analogous to the conventional expression of normalized bit rate in (5.6). Since, in general, the power spectrum of a scheme depends on shaping, ν must be determined for every choice of Ω and Φ.

5.5   Gain versus a Baseline Constellation

Using the definition of effective dimension ν (5.8) and the effective normalized rate (5.9), the optical power gain in (5.4) can be simplified to

$$G(\Omega, \nu) = \frac{|\Omega|^{1/\nu}}{2} \sqrt{\nu} \frac{d_{\min}(\Omega)}{P^G(\Omega)} \left(1 - |\Omega|^{-1/\nu} \right). \tag{5.10}$$

Note that the gain is independent of the value of T and $T_\oplus$, as in the conventional case. However, in the optical intensity case, the gain depends on the effective dimension ν as opposed to the dimension of the Euclidean space, N, as in the conventional case. For $|\Omega|$ large, or more precisely for a large effective normalized rates, the term $(1 - |\Omega|^{-1/\nu}) \approx 1$ and can be neglected.

5.6   Continuous Approximation to Optical Power Gain

The *continuous approximation* [112] replaces discrete sums of a function evaluated at every $x \in \Omega$ by normalized integrals of the function over the region $\Upsilon \cap \Psi$. For $f : \mathbb{R}^N \to \mathbb{R}$ which is Riemann-integrable over $\Upsilon \cap \Psi$ [109],

$$\sum_{x \in \Omega} f(x) \approx \frac{1}{V(\Lambda)} \int_{\Upsilon \cap \Psi} f(x) \, dV(x). \tag{5.11}$$

This approximation is valid when $V(\Upsilon \cap \Psi) \gg V(\Lambda)$, where the notation $V(\cdot)$ evaluates to the volume of the region. In practical terms, this condition holds when operating at a high effective normalized rates.

By the continuous approximation, $P^G(\Omega)$ (4.3) is approximated as

$$P^G(\Upsilon \cap \Psi) \approx \int_{x \in \Upsilon \cap \Psi} x_1 \frac{1}{V(\Upsilon \cap \Psi)} dV(x). \qquad (5.12)$$

Similarly, $|\Omega|$ is approximately

$$|\Omega| \approx \frac{V(\Upsilon \cap \Psi)}{V(\Lambda)}.$$

Since the continuous approximation is valid when $|\Omega|$ is large, in (5.10) the term $(1 - |\Omega|^{-1/\nu}) \approx 1$.

Since $d_{\min}(\Omega) = d_{\min}(\Lambda)$, (5.10) is well approximated by

$$G(\Omega, \nu) \approx \gamma_c(\Lambda, \nu) \gamma_s(\Upsilon, \Psi, \nu) \qquad (5.13)$$

where the *coding gain* is given as

$$\gamma_c(\Lambda, \nu) = \frac{d_{\min}(\Lambda)}{V(\Lambda)^{1/\nu}} \qquad (5.14)$$

and the *shaping gain* is

$$\gamma_s(\Upsilon, \Psi, \nu) = \frac{\sqrt{\nu}}{2} \frac{V(\Upsilon \cap \Psi)^{1/\nu}}{P^G(\Upsilon \cap \Psi)}. \qquad (5.15)$$

5.7 Coding Gain

In electrical channels, the coding gain is measured by Hermite's parameter introduced in (5.1). The electrical coding gain depends solely on the geometric properties of the lattice. In optical intensity channels, the coding gain(5.14) can be written in terms of $\gamma_n(\Lambda)$ as

$$\gamma_c(\Lambda, \nu) = d_{\min} \cdot \left(\sqrt{\gamma_n(\Lambda)}/d_{\min} \right)^{N/\nu}. \qquad (5.16)$$

Through ν, the optical coding gain depends on Φ, $\Upsilon \cap \Psi$ and Λ. Thus, the densest lattice in N-dimensions, as measured by $\gamma_n(\Lambda)$, may not maximize the optical coding gain in (5.14).

If transmitting at high effective normalized rates, Appendix 5.A demonstrates that the continuous approximation can be used to yield an estimate of the effective dimension, ν^C, independent of Λ. For a given $\Upsilon \cap \Psi$ and Φ, substituting $\nu = \nu^C$ in (5.16) leaves $\gamma_n(\Lambda)$ as the only term dependent on the lattice chosen. As a result, for large $|\Omega|$, the densest lattice in N-dimensions which maximizes $\gamma_n(\Lambda)$ also maximizes the optical coding gain, $\gamma_c(\Lambda, \nu)$.

5.8   Shaping Gain

The purpose of shaping is to reduce the average optical power requirement of a scheme at a given rate. In electrical channels, the shaping gain depends on the geometry of the constellation. As is the case with $\gamma_c(\Lambda, \nu)$, the shaping gain for the optical intensity channel (5.15) is a function of the region $\Upsilon \cap \Psi$ as well as ν. In general, determining the Ψ which maximizes $\gamma_s(\Upsilon, \Psi, \nu)$ is difficult and will depend on the specific basis functions selected.

5.8.1   Optimal Shaping Regions

It is possible to approximate ν as independent of $\Upsilon \cap \Psi$ in certain cases. If the $x(t) \in X$ have $K\%$ fractional energy bandwidth which are all approximately the same and $K \rightarrow 1$, then every symbol occupies essentially the same bandwidth. In this case, the bandwidth of the scheme, and hence ν, will be independent of $\Upsilon \cap \Psi$. Raised-QAM is shown to satisfy this approximation in Section 5.12.

If ν is independent of $\Upsilon \cap \Psi$, then the rate depends on the volume $V(\Upsilon \cap \Psi)$. The optimal choice of Ψ, in the sense of average optical power, is one which for a given volume, or rate, minimizes the average optical power P^{G}. The optimum shaping region which maximizes shaping gain is the half-space

$$\Psi^*(r_{\max}) = \{(\psi_1, \psi_2, \ldots, \psi_N) \in \mathbb{R}^N : r_{\max} \geq 0, \ \psi_1 \in [0, r_{\max}]\}, \quad (5.17)$$

for some fixed $r_{\max} \geq 0$ so that the desired volume is achieved. The proof of this assertion rests on noting that all points with equal components in the ϕ_1-dimension have the same average optical power. For a given volume, i.e., rate, and Υ, the optimal shaping region can be formed in a greedy fashion by successively adjoining points of the smallest possible average optical energy until the volume is achieved. It is clear that the region in (5.17) will result. This result is different than the case of electrical channels where the N-sphere is the optimal shaping region in an average energy sense.

As mentioned in Chapter 2, wireless optical channels are typically dominated by the average and not the peak optical power. However, in practical systems both constraints must be met. For a given pulse set, peak optical power $p/\sqrt{T}$ and rate, the optimum region which maximizes shaping gain is

$$\Psi^*(r_{\max}, p) = \Pi(p) \cap \Psi^*(r_{\max}), \quad (5.18)$$

for $\Psi^*(r_{\max})$ (5.17) and $\Pi(p)$ as defined in (4.7). The form of this region can be justified in an identical manner as the previous case, except that here the points selected such that they satisfy both non-negativity and peak amplitude constraints. Figure 5.4 presents an example of such a region for raised-QAM defined in Section 4.2.

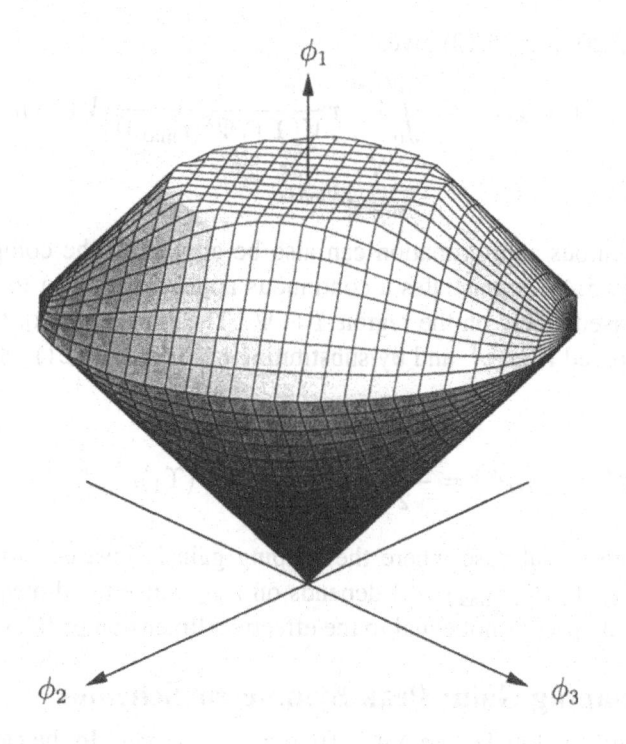

Figure 5.4. The region $\Psi^*(r_{\max}, p) = \Upsilon \cap \Psi^*(r_{\max}, p)$ for raised-QAM with $r_{\max} = (3/4)p$.

5.9 Shaping Gain: Expression

For constellations defined as $\Omega = \Lambda \cap \Upsilon \cap \Psi^*(r_{\max})$, the shaping gain in (5.15) can be simplified by exploiting the symmetries of Υ. By Property 1 of Theorem 4.1, the Υ_r are directly similar and scale linearly in r. As a result, the volume of each of the Υ_r scales as

$$V(\Upsilon_r) = V(\Upsilon_1)r^{N-1}, \tag{5.19}$$

for an N-dimensional signal space. The volume of $\Upsilon \cap \Psi^*(r_{\max})$ can then be computed simply as

$$
\begin{aligned}
V(\Upsilon \cap \Psi^*(r_{\max})) &= \int_0^{r_{\max}} V(\Upsilon_r)dr \\
&= \frac{1}{N}V(\Upsilon_1)r_{\max}^N.
\end{aligned}
\tag{5.20}
$$

The computation of average optical power (5.12) is simplified by exploiting the symmetry of the geometry and can be computed as an integral over the ϕ_1 direction only. Noting that $dV(\Upsilon \cap \Psi^*(r_{\max})) = V(\Upsilon_r)dr$ and substituting

(5.19) and (5.20) into (5.12) gives

$$P^G(\Upsilon \cap \Psi^*(r_{max})) = \int_0^{r_{max}} r \frac{1}{V(\Upsilon \cap \Psi^*(r_{max}))} (V(\Upsilon_1) r^{N-1}) dr$$

$$= \frac{N}{N+1} r_{max}. \tag{5.21}$$

The continuous approximation can also be applied to the computation of ν. Appendix 5.A demonstrates a continuous approximation to ν, ν^C, can be computed based solely on the region $\Upsilon \cap \Psi$. The expression for the shaping gain is computed from ν^C and by substituting (5.20) and (5.21) into (5.15) to yield

$$\gamma_s(\Upsilon, \Psi^*(r_{max}), \nu^C) = \frac{\sqrt{\nu^C}}{2} \left(\frac{(N+1)^{\nu^C}}{N^{\nu^C+1}} V(\Upsilon_1) r_{max}^{N-\nu^C} \right)^{1/\nu^C} \tag{5.22}$$

Unlike the electrical case where the shaping gain is invariant to scaling of the region, $\gamma_s(\Upsilon, \Psi^*(r_{max}), \nu^C)$ depends on r_{max} since the dimension of the defined signal space is not equal to the effective dimension of (Ω, Φ).

5.10   Shaping Gain: Peak-Symmetric Schemes

As demonstrated in Theorem 4.2, $\Pi(p) = -\Upsilon + p\phi_1$. In the case of peak-symmetric schemes, defined in Section 4.1.5, since $\text{Proj}(\partial \Upsilon_1)$ is closed under inversion $\Pi(p)$ is a ϕ_1-shifted *reflection* of Υ in the hyperplane $\phi_1 = 0$. As a result of this additional degree of symmetry, the cross-sections of Υ and $\Pi(p)$ in the ϕ_1-axis coincide for $\phi_1 = p/2$. In other words, the cross-sections of $\Upsilon \cap \Psi^*(r_{max}, p)$ for a given ϕ_1 value are all directly similar to Υ_1. Figure 5.4 presents the region $\Upsilon \cap \Psi^*(r_{max}, p)$ for a peak-symmetric, raised-QAM example defined in Section 4.2.

For peak-symmetric pulse sets with $r_{max} < p/2$, $\Upsilon \cap \Psi^*(r_{max}, p) = \Upsilon \cap \Psi^*(r_{max})$. The peak-symmetry of the scheme requires that all points in $\partial \Upsilon_{rmax}$ have a maximum amplitude of $2r_{max}/\sqrt{T}$ by (4.14). Thus, for $r_{max} < p/2$, all points in $\Upsilon \cap \Psi^*(r_{max})$ have peak amplitude less than p.

In the case of $r_{max} \in [p/2, p]$ the volume and P^G of the resulting region are,

$$V(\Upsilon \cap \Psi^*(r_{max}, p)) = \frac{1}{N} V(\Upsilon_1) \left(2 \left(\frac{p}{2} \right)^N - (p - r_{max})^N \right), \tag{5.23}$$

and

$$P^G(\Upsilon \cap \Psi^*(r_{max}, p)) = \frac{N}{N+1} \cdot \frac{(p/2)^N}{2(p/2)^N - (p - r_{max})^N}$$

$$\left(p \frac{N+1}{N} - \left(\frac{p - r_{max}}{p/2} \right)^N \left(\frac{1}{N} p + r_{max} \right) \right). \tag{5.24}$$

Substituting into (5.15) gives the shaping gain for peak-constrained regions. Note also that peak optical power in the case of these regions is $p/\sqrt{T}$ and gives a PAR of

$$\text{PAR}(\Upsilon \cap \Psi^*(r_{\max}, p)) = \frac{p}{P^G(\Omega)},$$

which is independent of the symbol interval.

5.11 Opportunistic Secondary Channels

Opportunistic secondary channels arise when there exists a degree of freedom in the selection of a constellation point. Say it is required to transmit b bits of data per symbol. If $|\Omega| > 2^b$ then there is a choice as to which signal point to transmit. By the proper exploitation of this choice it is possible to transmit data on this secondary channel. This increase in the size of the constellation will normally come at the expense of increased average energy [112].

For optical intensity lattice codes, this secondary channel can be exploited without increasing the average optical power. For the optimal shaping region $\Upsilon \cap \Psi^*(r_{\max})$, the maximum optical power shell $\Upsilon_{r\max}$ is in general not completely filled. All the lattice points in $\Upsilon_{r\max}$ are equivalent from an average optical power cost and the additional points in this shell can be selected without impact on the optical power of the scheme. This degree of freedom in selecting constellation points can be used to transmit additional data, introduce spectral shaping or add a tone to the transmit spectrum for timing recovery purposes.

5.12 Example Lattice Codes

5.12.1 Gain of Example Codes

The optical power gain versus the bandwidth efficiency is plotted for PPM and raised-QAM modulation schemes in Figure 5.5.

Ten discrete raised-QAM constellations were constructed as

$$\Omega = \mathbb{Z}^3 \cap \Upsilon \cap \Psi^*(r_{\max})$$

by selecting the appropriate $r_{\max}$ to have each carrying from 1 to 10 bits per symbol. The power spectral density for each scheme is computed symbolically via (3.14) using a symbolic mathematics software package [123] and integrated numerically to determine $W_{0.99}$ and $W_{0.999}$ for a fixed T. These results were then used to computed κ for each (Ω, Φ). The power spectral density of the baseline scheme is trivial to compute, and numerical integration yields $\kappa_\oplus = 20.572$ for $W_{0.99}$ and $\kappa_\oplus = 202.217$ for $W_{0.999}$. The effective dimension of the constellations are presented in Table 5.1. The same procedure was repeated for discrete PPM constellations of size 2 through 8. Table 5.2 presents the ν values computed for these constellations.

For both the raised-QAM and PPM schemes, ν in Tables 5.1 and 5.2 are essentially independent of the value of K in W_K. This suggests that ν is not

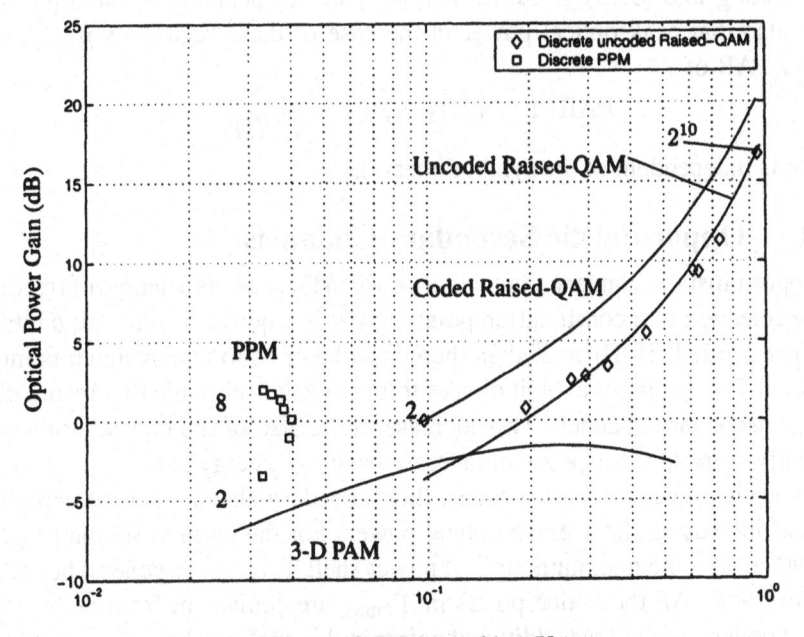

Figure 5.5. Gain over baseline versus bandwidth efficiency. Note that points indicated with □ and ◇ represent discrete PPM and raised-QAM constellations respectively while the solid lines represent results using the continuous approximation.

Table 5.1. Effective dimension for optimally shaped, discrete raised-QAM constellations.

$|\Omega|$	$W_{0.99}$ Bandwidth ν	$W_{0.999}$ Bandwidth ν
continuous	$\nu^c = 1.006$	$\nu^c = 1.000$
2	1.000	1.000
4	1.004	1.000
8	1.107	1.108
16	1.335	1.365
32	1.428	1.458
64	1.319	1.338
128	1.114	1.119
256	1.224	1.245
512	1.194	1.189
1024	1.024	1.033

Table 5.2. Effective dimension for discrete PPM constellations.

$|\Omega|$	$W_{0.99}$ Bandwidth ν	$W_{0.999}$ Bandwidth ν
2	2.990	2.994
3	3.939	4.002
4	4.893	4.993
5	5.985	5.993
6	6.810	7.002
7	7.855	7.984
8	8.907	8.982

sensitive to the choice of K for values of $K \to 1$. Note also that ν increases as $|\Omega|$ for the PPM constellations since each signal point is orthogonal to all others. In the case of raised-QAM, the effective dimension remains approximately constant as Ω increases.

The gain of the raised-QAM and PPM examples was computed using (5.10) and the ν derived from the $W_{0.99}$. Note that the raised-QAM schemes in Figure 5.5 provide large optical power gain over the baseline scheme while operating at high bandwidth efficiencies, while the PPM schemes provide small gain at low bandwidth efficiencies.

Using the optical power gain as a figure of merit can be deceptive since it depends on the baseline scheme chosen. The raised-QAM and PPM examples operate at different bandwidth efficiencies and a direct comparison of their performance is not possible using this measure. Indeed, Figure 5.5 suggests that PPM and raised-QAM are suited for operation under highly different channel conditions. In Section 5.12.2 a comparison technique based on an idealized point-to-point link is presents which illustrates the conditions under which PPM or raised-QAM are appropriate.

It is often difficult if not impossible to directly compute the power spectral density to find ν. To verify the asymptotic accuracy of the continuous approximation gain in (5.13), ν^C was computed for the raised-QAM examples. The continuous approximation to the psd was computed symbolically using (5.A.1) and then integrated numerically to get an estimate of ν^C. Table 5.1 presents the estimated value of effective dimension. The gain using the continuous approximation for ν is included in Figure 5.5 and demonstrate that for large constellations that the continuous approximation to the optical power gain approaches the discrete case.

For the 3-dimensional PAM bases, defined in Section 4.2.1, the continuous approximation of gain was computed in an identical fashion to give $\nu^C = 3.629$

and is also plotted in Figure 5.5. It is somewhat surprising that the baseline is more power efficient than the 3-D PAM scheme in spite of the fact that the 3-D PAM constellation arises as a shaped version of the baseline using $\Psi^*(r_{max})$. This is due to the fact that ν depends on $\Upsilon \cap \Psi$. The effective dimension for the shaped case is larger than the baseline value of 3 which eliminates any shaping gain. The approximation that the bandwidth of each symbol is approximately constant is not good for the 3-D PAM example, and thus the interpretation of $\Psi^*(r_{max})$ as optimal in average optical power at a given rate is not true.

The role of coding is investigated by forming a 24-dimensional constellation by specifying symbols consisting of blocks of 8 consecutive 3-dimensional raised-QAM symbols. The resulting Υ was intersected with the Leech lattice, Λ_{24}, to form a constellation. Using the continuous approximation, the optical power gain was calculated. The effective dimension was approximated by integrating over the 24-dimensional region symbolically to give $\nu^C = 8.268$. The optical gain using this coding is plotted in Figure 5.5 for comparison. The use of Λ_{24} over $\mathbb{Z}^{24}$ gives a coding gain of approximately 3 dB in optical power which is less than the 6 dB nominal electrical coding gain [120]. This reduction is coding gain arises since, as alluded to in (5.16), the optical coding gain depends on the square root of Hermite's parameter. Qualitatively, an electrical coding gain corresponds to a reduction in the mean square value of the signal while an optical coding gain corresponds to the reduction in the mean value of the signal.

5.12.2   Idealized Point-to-Point Link

Using the gain over a baseline as a metric for comparison on modulation schemes can often be misleading. The optical power gain is highly dependent on the baseline scheme that is chosen. In order to have a more consistent comparison between PPM and raised-QAM schemes, the distance over which scheme could transmit over an idealized eye-safe, point-to-point wireless optical channel operating at a given symbol error rate and data rate was taken as a measure of optical power efficiency. Figure 5.6 presents a block diagram of the idealized link. The receiver and transmitter are assumed to be aligned and a distance D cm apart. The average transmitter intensity is constrained to $\bar{I} = 104$ mW/steradian, which is the eye-safety limit of a commercially available wireless infrared transceiver [124]. The detector sensitivity is taken to be $r = 25\,\mu\text{A} \cdot \text{m}^2/\text{mW}$ over an area of 1 cm^2 and the channel noise standard deviation $\sigma = 11.5 \times 10^{-12}\,\sqrt{\text{W/Hz}}$ both of which have been reported for a similar experimental link [125]. The symbol error rate in all cases was set to 10^{-8}, which corresponds to the IrDA Fast IR specification [85]. Assuming operation in the far-field case, the transmission distance under these constraints

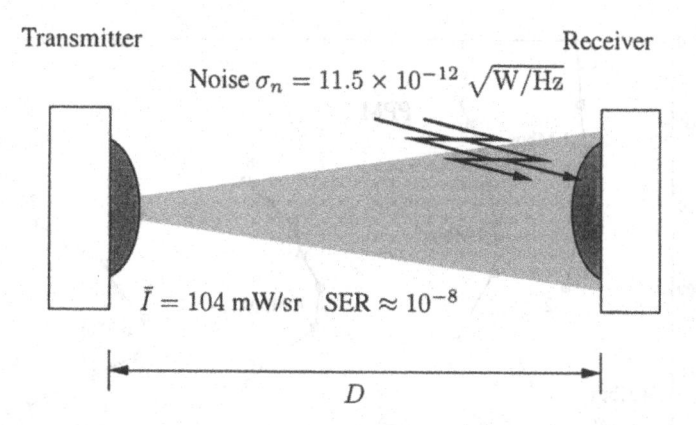

Transmitter   Receiver

Noise $\sigma_n = 11.5 \times 10^{-12} \sqrt{\text{W/Hz}}$

$\bar{I} = 104$ mW/sr   SER $\approx 10^{-8}$

D

Figure 5.6.   Block diagram of an idealized point-to-point link.

is

$$D = \sqrt{\frac{r\bar{I} \cdot \text{CFM}(\Omega, \nu)}{2k\sigma}},$$

where $k \approx 5.6120$, which is set by the symbol error rate.

Figure 5.7 presents the distance of the point-to-point link versus bandwidth for the discrete uncoded examples presented in Section 5.12.1 and Figure 5.5 with a fixed symbol error rate and for a variety of data rates. The bandwidth measure is $W_{0.99}$. For a given constellation, the data rate is varied by varying T. For both schemes increasing the rate for a given constellation causes a reduction in the distance. This reduction in range arises due to the fact that P (4.3) is inversely proportional to $\sqrt{T}$. Increasing the data rate by decreasing T for a given constellation then increases the optical power required to achieve the given symbol error rate and hence reduces transmission range. As discussed in Section 4.1.1 this is due to the unit energy normalization of basis functions in the signal space.

The idealized point-to-point link also indicates that pulse position modulation schemes are appropriate for long-range transmission at the price of increased bandwidth while raised-QAM schemes are appropriate for high data rate, short distance links. Long-range links are limited by the amount of power which can be collected at the receiver. As a result, power efficient PPM scheme must be employed at the cost of bandwidth expansion or equivalently rate loss. Small cardinality PPM signalling sets (of size 2 or 3) have limited use since the optical power efficiency is small compared to the required bandwidth. Figure 5.7 demonstrates that PPM constellations of cardinality 4 or more allow for a trade-off between rate and link distance. As the link distance becomes smaller, the amount of power collected increases. In this case, bandwidth efficient, multi-level, raised-QAM techniques can be employed to increase the link data

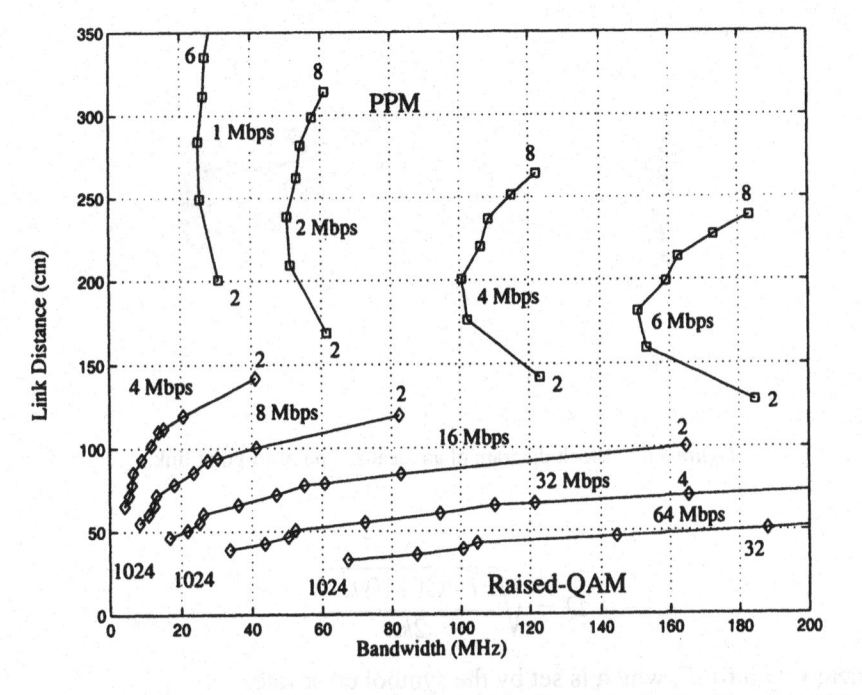

Figure 5.7. Idealized point-to-point link length versus bandwidth for the discrete uncoded PPM (□) and raised-QAM (◊) constellations of Figure 5.5 with SER $\approx 10^{-8}$.

rate. The idealized point-to-point link comparison technique quantifies the rate versus link distance trade-off in the design of wireless optical links.

The results of the idealized point-to-point link in Figure 5.7 also provide a design guide for the construction of modulation schemes. If we assume that an industry standard IrDA Fast IR link using 4-PPM and operating at 4 Mbps is taken as the operating point, using the 99% fractional power bandwidth, the link bandwidth is 100 MHz. For the sake of comparison, the first-null bandwidth for rectangular PAM in this case is approximately 9.7 MHz. This link is able to support a data rate of 4 Mbps at the specified 10^{-8} symbol error rate. If the same physical link is required to transmit at 125 cm, a 2-raised-QAM scheme can be used to achieve rates of 8 Mbps. If the distance is reduced to 75 cm, rates of up to 16 Mbps are possible using 4-raised-QAM. Data rates of 32 Mbps over 60 cm using 32-raised-QAM and 64 Mbps over 40 cm using 256-raised-QAM constellations are possible. Thus, for a given set of optoelectronics, which determine the channel bandwidth, the data rate can be optimized by the proper selection of modulation scheme. Power-efficient schemes operate over longer distances at low data rates while bandwidth efficient schemes offer high data rates over shorter distances.

Note that for raised-QAM in Figure 5.5 the gain of increases with $|\Omega|$ while in Figure 5.7 the link distance decreases with $|\Omega|$. This is because the optical

Table 5.3. Effective dimension for raised-QAM constellations shaped by $\Psi^*(r_{max}, p)$.

r_{max}/p	$W_{0.99}$ Bandwidth ν^C	$W_{0.999}$ Bandwidth ν^C
0.50	1.006	1.000
0.55	1.006	1.000
0.60	1.005	1.000
0.65	1.005	1.000
0.70	1.005	1.000
0.75	1.005	1.000
0.80	1.005	1.000
0.85	1.004	1.000
0.90	1.004	1.000
0.95	1.004	1.000
1.00	1.004	1.000

power requirement of the baseline scheme increases more quickly than raised-QAM at high bandwidth efficiencies. The use of link distance is a more practical means of computing relative performance of signalling schemes.

5.12.3 Peak Optical Power

The peak optical power of the example schemes was thus far not considered. Here, the impact of shaping with $\Psi^*(r_{max}, p)$ on fractional power bandwidth and on average optical power are investigated.

The ν^C values for raised-QAM constellations shaped with $\Psi^*(r_{max}, p)$ are presented in Table 5.3. For a given r_{max} and p, scaling of the region $\Psi^*(r_{max}, p)$ does not alter ν^C and so the set of regions can be parameterized by $k = r_{max}/p$. It can be shown that the symbols are all nearly bandlimited in the sense of Section 5.8.1, the effective dimension is approximated as being nearly independent of $\Upsilon \cap \Psi$. Thus, for the raised-QAM example and under the assumptions stated in Section 5.8.1, $\Psi^*(r_{max})$ and $\Psi^*(r_{max}, p)$ are optimal in the average optical power sense.

At a given rate, the use of shaping region $\Psi^*(r_{max}, p)$ reduces the peak optical power of a scheme at the cost of increasing average power over the case using shaping region $\Psi^*(r_{max})$. For the raised-QAM bases, form the constellations $\Omega_1 = \Lambda \cap \Upsilon \cap \Psi^*(r_{max})$ and $\Omega_2 = \Lambda \cap \Upsilon \cap \Psi^*(r'_{max}, p)$, for some fixed r_{max} and $k = r'_{max}/p$ for $k \in [0.5, 1]$. Under the assumption that ν is approximately independent of Ψ, fixing $V(\Upsilon \cap \Psi^*(r_{max})) = V(\Upsilon \cap \Psi^*(r'_{max}, p))$, defined in (5.20) and (5.23), fixes the rates of the schemes to be equal. The peak constraint incurs an excess average optical power penalty,

$$\frac{P(\Upsilon \cap \Psi^*(kp, p))}{P(\Upsilon \cap \Psi^*(r_{\max}))},$$

which can be computed via (5.21) and (5.24). The normalized peak optical power of Ω_2 with respect to Ω_1 is defined as

$$\frac{p}{2r_{\max}},$$

where the peak optical power of Ω_1 is defined in (4.14).

The trade-off between the peak and average optical power at a given rate for the uncoded raised-QAM example is presented in Figure 5.8. The independence of ν and Ψ is illustrated in the figure since results using the approximation and results in which ν was computed numerically nearly coincide. Additionally, Figure 5.8 demonstrates that the peak value of the constellation can be reduced by approximately 1 dB over the case of $\Psi^*(r_{\max})$ at a cost of less than 0.25 dB increase in average optical power. It is often important to reduce peak amplitudes in practical systems since schemes with high peak optical amplitudes is more vulnerable to channel nonlinearities and requires more complex modulation circuitry.

5.13   Conclusions

The signalling design problem for wireless optical intensity channels is significantly different than for the conventional electrical channel. Whereas in electrical channels the constraint is on the mean square value of the transmitted signal, the optical intensity channel imposes the constraint that all signals are non-negative and that the average signal amplitude is limited. Here a further constraint is imposed on the bandwidth of the channel since wireless optical channels are typically bandwidth constrained. A further constraint on the peak optical power is also considered and the trade-off between average and peak optical power is quantified.

Popular PPM schemes provide a means to trade-off optical power efficiency for bandwidth efficiency. In point-to-point links, this manifests itself as a trade-off between transmission distance and data rate. Bandwidth efficient multi-level schemes, such as raised-QAM, provide higher data rates at the expense of a greater required optical power. In this chapter, the linearity of the transmit and receive optoelectronics is not considered. Indeed, in practical links the non-linearity, especially in some laser diodes, may limit the implementation of multi-level modulation schemes.

It must be emphasized that the use of optical power gain versus the baseline is only useful when comparing two schemes at a given bandwidth efficiency. This technique was used to determine the optical power coding gain for a 24-dimensional Leech lattice example. To compare modulation schemes with

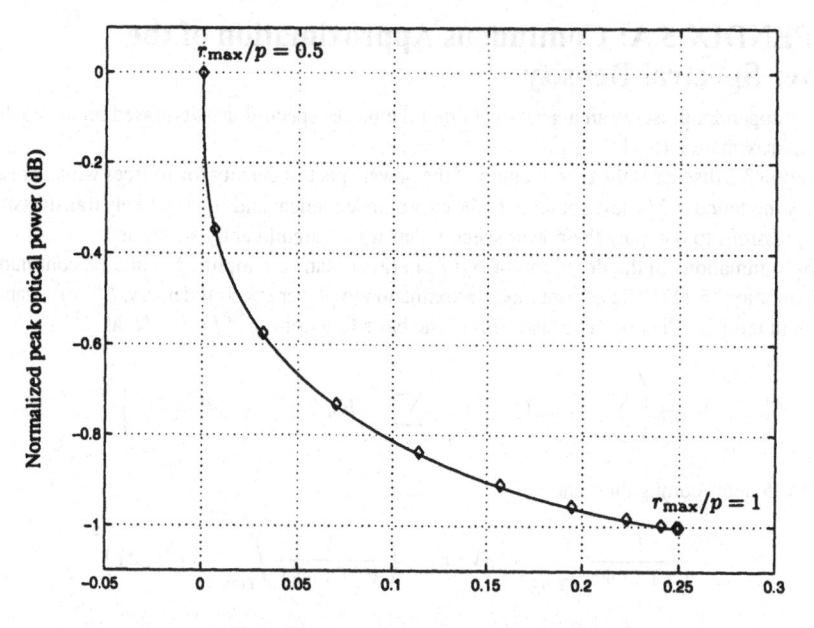

Figure 5.8.   Peak optical power versus excess average optical power (at same rate). Solid line represents case where ν is assumed independent of Ψ, and points $\Diamond$ indicate values for which ν^C, in Table 5.3, was computed numerically.

widely differing bandwidth efficiencies the baseline comparison method is not longer effective and comparisons must using idealized point-to-point link.

This chapter has further demonstrated that coding alone, although necessary to approach capacity, provides relatively limited optical power gain. The optical coding gain is shown to be proportional to the square root of the electrical coding gain. In the design of modems for such a channel, physical improvements to improve the optical power efficiency should first be exploited before complex coding schemes are considered. Chapter 9 briefly discusses the use of optical concentrators and long wavelength devices increases the receive optical SNR and allows for the use of high-data rate spectrally efficient modulation.

Although a number of example pulse sets and coding schemes are treated in this chapter, the question of the maximum achievable data rate for a given pulse set remains open. Chapter 6 answers this question by computing bounds on the capacity of optical intensity pulse sets subject to an average optical power constraint.

APPENDIX 5.A: Continuous Approximation of the Power Spectral Density

This appendix presents an approximation of the power spectral density based on the continuous approximation (5.11).

Section 3.2 discusses the computation of the power spectral density for sources which are accurately modelled as Markov sources. In the case of independent and equally likely transmission, the expressions to compute the power spectral density are significantly simplified.

The summations in the definition of $S_X^c(f)$ in (3.14) can be simplified using the continuous approximation (5.11). The continuous approximation to power spectral density, $S_X^c(f)^C$, can be written in terms of the Fourier transforms of the basis functions, $\phi_i^F(f)$, $i \in N$, as

$$S_X^c(f)^C = \frac{1}{T}\left(\sum_{n \in N} J_n |\phi_n^F(f)|^2 + \sum_{n,l \in N,\ n \neq l} K_{nl}\mathrm{Re}\{\phi_n^F(f)\phi_l^{F*}(f)\}\right) \tag{5.A.1}$$

where $*$ denotes conjugation and

$$J_n = \frac{1}{V(\Upsilon \cap \Psi)}\int_{\Upsilon \cap \Psi} x_n^2 dV(x) - \left(\frac{1}{V(\Upsilon \cap \Psi)}\int_{\Upsilon \cap \Psi} x_n dV(x)\right)^2,$$

$$K_{nl} = \frac{1}{V(\Upsilon \cap \Psi)}\int_{\Upsilon \cap \Psi} x_n x_l dV(x) - \frac{1}{V^2(\Upsilon \cap \Psi)}\int_{\Upsilon \cap \Psi} x_n dV(x)\int_{\Upsilon \cap \Psi} x_l dV(x).$$

Therefore, in order to calculate the continuous approximation for the power spectral density the first and all second order moments of the N-dimensional random vector uniformly distributed over $\Upsilon \cap \Psi$ must be determined.

Figure 5.A.1 plots $S_X^c(f)^C$ along with the power spectral density calculated symbolically via (3.14) for various sizes of discrete, optimally shaped, uncoded raised-QAM examples in Section 5.12. As the rate increases, it is apparent that $S_X^c(f)^C$ approaches the true power spectral density (3.14).

An approximation to ν in (5.8) can be computed by estimating the fractional power bandwidth of a scheme via numerical integration of (5.A.1) to yield ν^C. The accuracy of this approximation improves of the rate of the scheme increases, as is evident in Figure 5.A.1.

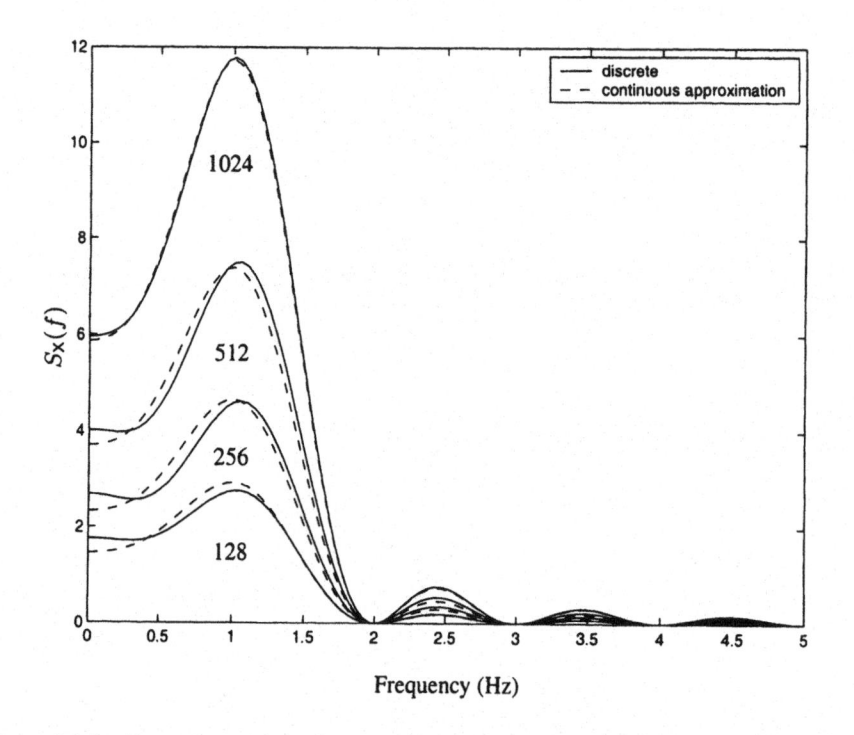

Figure 5.A.1.   Power spectral density of uncoded, discrete raised-QAM constellations (solid line) versus the continuous approximation for the power spectral density (dashed line, $T = 1$ in all cases).

Chapter 6

CHANNEL CAPACITY

There has been much work in the investigation of the capacity of optical channels in which the dominant noise source is quantum in nature. These Poisson photon counting channels, discussed in Section 3.1.5, are good models when the receiver is illuminated with low intensity light. Wireless optical channels, as described in Chapter 2, are subjected to intense background illumination which dominates the quantum noise in the channel.

This chapter investigates the fundamental limits on communication in wireless optical channels [126]. Asymptotically exact upper and lower bounds on the capacity of optical intensity channels are presented on the vector additive white Gaussian noise channel. These capacity bounds are not restricted to rectangular pulse sets, but use the signal space model of Chapter 4 to represent all time-disjoint schemes. The chapter concludes with the derivation of capacity bounds for a number of basis sets and concludes that spectrally efficient pulse selection is critical in obtaining high spectral efficiencies.

6.1 Background

6.1.1 What is Channel Capacity?

The *channel capacity* is a fundamental measure of the maximum amount of information which can be conveyed through a channel reliably [127]. Shannon's discovery that it was possible to have arbitrarily reliable communication at non-zero rates revolutionized communication system design practices and established the areas of information theory and error control coding. The capacity, C, of a discrete memoryless vector channel from X to Y is

$$C = \sup_{f_X(x) \in \mathcal{F}} I(X;Y) \tag{6.1}$$

where $I(X, Y)$ is the mutual information between input and output vectors in a sample space and $f_X(x)$ is the source distribution [128]. The set $\mathcal{F}$ is the collection of all distributions that satisfy some cost constraint. A typical example of a cost constraint would be the set of all distributions with a given variance or electrical power constraint. For example, a Gaussian noise channel, with variance σ_n^2 where $\mathcal{F}$ is the set of all distributions on the real line with variance at most σ_p^2, has capacity,

$$C = \frac{1}{2} \log_2 \left(1 + \frac{\sigma_p^2}{\sigma_n^2} \right), \tag{6.2}$$

where the capacity achieving source distribution is normal with zero mean and variance σ_p^2. Shannon proved in his *channel coding theorem* [127] that a necessary and sufficient condition for the existence of a coding scheme which can operate at an arbitrarily low probability of decoding error is that the rate of information transmission is no larger than the channel capacity. Thus, the capacity of a channel is a fundamental measure of performance and represents the ultimate rate at which information can be conveyed *reliably*. There any many excellent references on information theory which outline the basic concepts and practices in greater detail [128–130].

6.1.2   Previous Work

Historically, the optical intensity channel has been modelled as a Poisson counting channel in which photon counts at the receiver are corrupted by the quantum nature of the channel as well as by external noise sources. In the absence of external noise, Gordon and Pierce demonstrated that the capacity of such photon counting channels (in units of nats/photon) under an average optical power limit is unbounded [131, 132]. In fact, M-ary PPM can achieve arbitrarily small probability of error for any rate [132, 113]. Under an additional constraint of fixed peak optical power, Davis and Wyner determined the capacity of this channel and showed that binary level modulation schemes were capacity-achieving [133, 134]. These channel capacity results, however, do not impose an explicit bandwidth constraint on the signals transmitted. Indeed, McEliece demonstrated that schemes based on photon counting in discrete intervals require an exponential increase in bandwidth as a function of the rate (in nats/photon) for reliable communication [113]. In the case of rectangular PAM signals confined to discrete time intervals of length T and with a given peak and average optical power, Shamai showed that the capacity achieving input distribution is discrete with a finite number of levels increasing with T [135]. For high bandwidth cases, as $T \to 0$, the binary level techniques found earlier are capacity achieving, however, lower bandwidth schemes require a larger number of levels. In the case of intense ambient lighting, where the noise can be modelled as being Gaussian distributed, it has been demonstrated that the capacity achieving distribution for multi-dimensional signalling subject to a peak and average optical power constraints is always in some sense "discrete", i.e., is defined on a set of measure zero [136]. It still remains difficult, however, to find the capacity achieving distribution in all but the case of one dimensional signalling.

The maximum rate of multiple PPM on photon counting channels was computed for channels with no inter-symbol interference [92]. In the case of multipath wireless optical channels, the maximum rate for MPPM was computed in the presence of inter-symbol interference [137]. In the case of multiple subcarrier modulation for wireless optical channels, an upper bound on the channel capacity was computed subject to an average optical power constraint under the assumption that the average optical power per symbol was fixed [138]. The bounds derived in this chapter differ significantly in that they do not assume a pulse shape and they allow the average optical amplitude to vary from symbol to symbol.

6.2 Problem Definition

As in Chapters 4 and 5, consider transmitting symbols formed as linear combinations of the N basis functions in the set Φ. With N matched filters at the receiver, the channel can be modelled as the vector channel (3.5),

$$Y = X + Z$$

where each term is an N-dimensional random vector with probability densities $f_Y(y)$, $f_X(x)$ and $f_Z(z)$ respectively and Z is an N-dimensional Gaussian random vector with independent components of mean zero and variance σ^2 per dimension.

The transmitted vector X must represent a signal which satisfies the amplitude constraints of the optical intensity channel. The set of source distributions, $\mathcal{F}$ in (6.1) must contain distributions which have support only for those values which correspond to non-negative output amplitudes and satisfy average (and perhaps peak) amplitude constraints. That is, in the signal space introduced in Chapter 4.1.1, all of the distributions in $\mathcal{F}$ have support on Υ. Explicitly, for $f_X(x) \in \mathcal{F}$, in order to satisfy the non-negativity constraint, $f_X(x) = 0$ for $x \notin \Upsilon$. To the knowledge of the author, no closed form solution for the capacity of wireless optical intensity channel which are modelled in this manner.

The average optical power constraint requires that for every $f_X(x) \in \mathcal{F}$,

$$
\begin{aligned}
P &\geq \frac{1}{\sqrt{T}} \int_{x \in \Upsilon} x_1 f_X(x) dx \\
&= \frac{1}{\sqrt{T}} P^G,
\end{aligned}
\tag{6.3}
$$

for a given bounded value of P. Note that this is analogous to the definition in (4.3) except that $f_{\boldsymbol{X}}(\boldsymbol{x})$ need not be discrete as was the case in Chapter 5.

The problem of finding the maximum rate at which reliable communication can take place in the optical intensity case is

$$C_{\mathrm{s}}(\Phi) = \max_{f_{\boldsymbol{X}}(\boldsymbol{x}) \in \mathcal{F}} I(\boldsymbol{X}; \boldsymbol{Y}) \qquad (6.4)$$

in units of bits per channel use were $\mathcal{F}$ is the set of source distributions satisfying the amplitude constraints. This capacity is the maximum achievable rate when the transmitted waveforms are restricted to be linear combinations of a particular basis set Φ. This result is analogous to the case of strictly bandlimited electrical channel where all transmitted signals are a scaled version of the basis function $\mathrm{sinc}(\pi t/T)$ and the channel capacity is

$$C = W \log_2(1 + \mathrm{SNR})$$

where W is the channel bandwidth and SNR is the signal-to-noise ratio [128].

This chapter finds bounds on the maximum rate at which reliable communication can take place under non-negativity and average optical power constraints for a family of channel models determined by the choice of Φ.

6.3   Bandwidth Constraint

Capacity results on photon counting channels indicated that the rate is unbounded if the average optical power is the only constraint [131, 132]. It is possible to show that arbitrarily narrow rectangular pulses in a symbol interval of T at a given average optical power causes the capacity in (6.4) to grow without bound. However, previous work also noted that this unbounded rate necessarily comes at the cost of an infinite bandwidth and amplitude requirement [113]. Therefore, it is necessary to specify a bandwidth constraint on the set of transmitted signals in order to have a consistent notion of channel capacity.

There any many notions of bandwidth, some of which are discussed in Section 3.2. In the previous chapter on lattice codes, Section 5.4, a bandwidth constraint on a constellation was interpreted as an effective dimension due to an analogy with electrical bandlimited modem design. Unlike the case treated in Chapter 5, where specific signal sets were considered, the capacity computation is a maximization over all possible signalling sets.

In this case, a bandwidth constraint is imposed on the set of transmitted signals by way of the Landau-Pollak dimension, presented in Section 3.2. It was shown, for an energy signal $x(t)$ with $(1 - \epsilon)$-fractional energy bandwidth of $W_\epsilon(x)$ and symbol period T that approximately $2W_\epsilon(x)T$ orthonormal prolate spheroidal wave functions were required to describe $x(t)$. An *effective dimension* of the signal space associated with optical intensity basis Φ can be defined as,

$$\kappa(\Phi) = \max_{x \in \Upsilon} 2W_\epsilon \left(\sum_{n \in \mathbb{N}} x_n \phi_n(t) \right) T. \tag{6.5}$$

This definition is equivalent to fixing T and specifying the available channel bandwidth, W_{ch}, as the largest bandwidth of any transmittable signal, i.e., $W_{ch} = \kappa(\Phi)/2T$. This channel bandwidth definition ensures that the channel is able to support the transmission of at most $\kappa(\Phi)$ dimensions per symbol. Since each transmitted waveform is at most $\kappa(\Phi)$ dimensional, the received signals are uncorrupted by the channel, i.e., on a noiseless channel, the received signals are indistinguishable from the transmitted signals in the sense of (3.18).

It is often difficult to compute $\kappa(\Phi)$ for a given basis set Φ. Practically, $\kappa(\Phi)$ is computed by fixing T and maximizing the bandwidth over all transmittable signals. Without loss of generality, consider an $x(t) \in L^2[0,T]$ which is unit energy. The bandwidth $W_\epsilon(x)$ can be approximated by expanding $x(t)$ in terms of the basis Φ as

$$W_\epsilon(x) \approx \inf \left\{ W \in [0,\infty) : \sum_{n \in \mathbb{N}} x_n^2 \int_{-W}^{W} |\phi_n^F(f)|^2 df \geq (1-\epsilon) \right\}, \tag{6.6}$$

where $\phi_n^F(t)$ is the Fourier transform of $\phi_n(t)$ and $\sum_m x_m^2 = 1$. This approximation holds for large W or equivalently for small ϵ, when the approximation

$$\int_{-W}^{W} \phi_m^F(f) \phi_n^{F*}(f) df \approx 0.$$

is valid for $m \neq n$. This approximation is used in several of the examples presented in Section 6.6.

6.4 Upper bound on Channel Capacity

An upper-bound on the capacity of the Gaussian-noise-corrupted wireless optical intensity channel can be obtained by considering a sphere-packing argument in the set of all received codewords while imposing the amplitude constraints of the channel. This analysis is done in the same spirit as Shannon's sphere packing argument for channels subject to an average electrical power constraint [139, 140]. To find this upper bound, the volume of the set of received codewords be computed for a given average optical power limit.

6.4.1 Background

Consider transmitting a codeword x formed from a series of L, N-dimensional symbols drawn from a constituent constellation. In order for x to be transmittable, $x \in \Upsilon^L$, where Υ^L is the L-fold Cartesian product of Υ with itself.

As discussed in Section 4.1, the admissible region, Υ, of any time-disjoint signalling scheme is the convex hull of a generalized cone. The Cartesian product Υ^L represents a set of signals which satisfy the non-negativity constraint and are time limited. Therefore, Υ^L represents a time-limited optical intensity scheme and must therefore be the convex hull of a generalized cone with vertex at the origin.

An LN-dimensional transmitted vector, x, is drawn from the set

$$\Theta(\sqrt{T}P) = \Upsilon^L \cap \Psi(P^G),$$

where the shaping region $\Psi(P^G)$ is a hyperplane defined to ensure that the power constraint (6.3) is satisfied. The region $\Psi(P^G)$ can be expressed in terms of the signal space as

$$\Psi(P^G) = \left\{ x \in \mathbb{R}^{LN} : \frac{1}{L} \sum_{k=1}^{L} x_{1,k} \leq P^G \right\}$$

where the $x_{1,k}$ are the ϕ_1 components of each of the L constituent constellations.

Given that $x \in \Theta(\sqrt{T}P)$ is transmitted, the received vector, Y, is normally distributed with independent components of variance σ^2 per dimension about mean vector x. Denote the set of all possible received vectors by Γ_{LN}. By the law of large numbers, with high probability Y will lie near the surface of a sphere of radius $\sqrt{LN(\sigma^2 + \epsilon)}$ where ϵ can be made arbitrarily small by increasing L [140]. A codeword is decoded by assigning all vectors contained inside the sphere to the given codeword.

Define Γ_∞ as

$$\Gamma_\infty = \{ x + b : x \in \Theta(\sqrt{T}P), \, b \in \rho B^{LN} \} \tag{6.7}$$

where

$$\rho = \sqrt{LN\sigma^2}, \tag{6.8}$$

and B^{LN} is the LN-dimensional unit ball. For all $y \in \Gamma_{LN}$ which arise from the transmission of $x \in \Theta(\sqrt{T}P)$ and any given positive ϵ and δ, the inequality

$$\Pr(| \|y - x\|^2 - \rho^2 | < \epsilon) \geq 1 - \delta$$

can be satisfied for sufficiently large L by the weak law of large numbers. Thus, for large enough L the distance between any $y \in \Gamma_{LN}$ and the corresponding $x \in \Theta(\sqrt{T}P)$ tends to ρ with probability arbitrarily close to one. In other words, for large enough L the probability that y does not lie in Γ_∞ can be made arbitrarily small. Since the capacity calculations depend on the asymptotic behavior of Γ_{LN} in L, the properties of Γ_∞ must be determined in order to determine an upper bound.

Figure 6.1 presents an example of a two-dimensional cross-section of $\Theta(\sqrt{T}P)$ and Γ_∞.

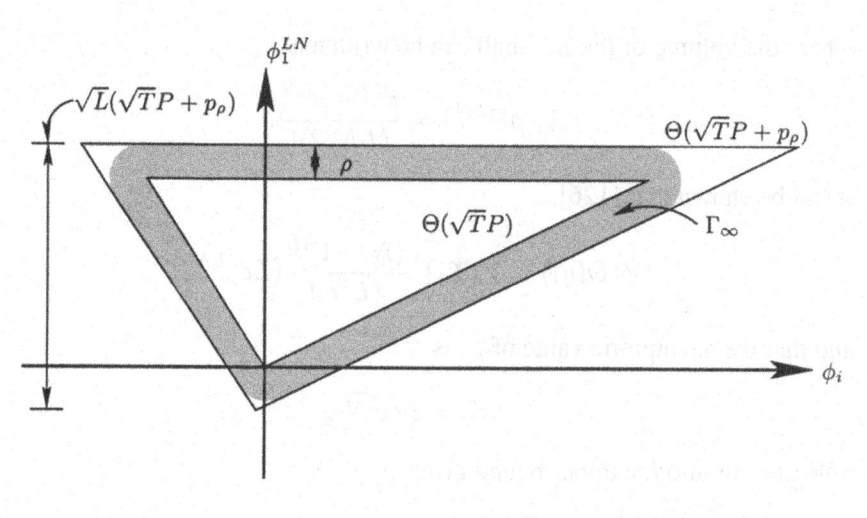

Figure 6.1. Cross-section of $\Theta(\sqrt{T}P)$, Γ_∞ and upper bound $\Theta(\sqrt{T}P + p_\rho)$.

6.4.2 Upper bound Computation

The channel capacity in bits/symbol for a given optical intensity basis set, $C_s(\Phi)$, can be upper bounded using the sphere packing argument developed for electrical power constrained channels [139]. The sphere-packing argument is used to upper bound channel capacity by finding an upper bound on the number of non-overlapping "message spheres" of radius ρ that can be place in Γ_{LN} as $L \to \infty$. The spheres are non-overlapping since, qualitatively, any overlap will yield confusion about which message was transmitted and arbitrarily low probability of error cannot be achieved. These concepts can be formalized rigorously in Gaussian channels [140].

Using the previously defined regions,

$$C_s(\Phi) \le \lim_{L \to \infty} \frac{1}{L} \log_2 \frac{V(\Gamma_{LN})}{V(\rho B^{LN})}$$

The computation of the volume of Γ_{LN} is in general very difficult due to the geometry of the region, however, it is possible to upper bound the region. It is possible to show [126] that

$$\Theta(\sqrt{T}P) \subset \Gamma_\infty \subset \Theta(\sqrt{T}P + p_\rho) - w. \tag{6.9}$$

for some constants p_ρ and w, as illustrated in Figure 6.1. Therefore, $C_s(\Phi)$ can be further upper bounded as

$$C_s(\Phi) \le \lim_{L \to \infty} \frac{1}{L} \log_2 \frac{V(\Theta(\sqrt{T}P + p_\rho))}{V(\rho B^{LN})} \tag{6.10}$$

where the volume of the LN-ball can be written as

$$V(\rho B^{LN}) = \frac{\pi^{LN/2}\rho^{LN}}{(LN/2)!}.$$

It can be shown that [126],

$$V(\Theta(q)) = V(\Upsilon_1)^L \frac{(N-1)!^L}{(LN)!}(Lq)^{LN},$$

and that the asymptotic value of p_ρ is

$$p_\rho = 2\sigma\sqrt{N}.$$

Substituting into the upper bound gives,

$$\frac{V(\Theta(\sqrt{T}P + p_\rho))}{V(\rho B^{LN})} = \frac{(LN/2)!}{(LN)!}\frac{\left((L(P^G + 2\sigma\sqrt{N}))^N V(\Upsilon_1)(N-1)!\right)^L}{(LN\pi\sigma^2)^{LN/2}}.$$

Using Stirling's formula [141] to bound the factorial functions,

$$\sqrt{2\pi n}\left(\frac{n}{e}\right)^n \exp\left(\frac{1}{12n+1}\right) < n! < \sqrt{2\pi n}\left(\frac{n}{e}\right)^n \exp\left(\frac{1}{12n}\right),$$

and simplifying the expressions, the number of codewords can be upper bounded as

$$\frac{V(\Theta(\sqrt{T}P + p_\rho))}{V(\rho B^{LN})} <$$

$$\frac{\exp f_\epsilon(L)}{\sqrt{2}}\left(\frac{(P^G + 2\sigma\sqrt{N})V(\Upsilon_1)^{1/N}(N-1)!^{1/N}}{N\sigma}\sqrt{\frac{e}{2\pi}}\right)^{LN}$$

where $\lim_{L\to\infty} f_\epsilon(L) = 0$.

Substituting into (6.10), replacing P^G with the average optical power through (6.3) and taking the limit as $L \to \infty$ yields the upper bound,

$$C_s(\Phi) \le N \log_2\left[\left(\sqrt{T}\frac{P}{\sigma} + 2\sqrt{N}\right)\frac{V(\Upsilon_1)^{1/N}(N-1)!^{1/N}}{N}\sqrt{\frac{e}{2\pi}}\right]$$

$$\text{[bits/symbol]} \quad (6.11)$$

for symbol period T. The upper bound depends on the pulse set chosen through the volume of the admissible region Υ. The selection of pulse shapes to maximize achievable rates is treated in Section 6.6.4 for a given bandwidth and over signal-to-noise ratio.

As discussed earlier, in order to have a consistent measure of maximum rate a spectral constraint must be imposed on the set of signals. The capacity can be expressed as a maximum spectral efficiency in units of bits/second/Hertz, $C_\eta(\Phi)$, as opposed to $C_s(\Phi)$ in (6.11), which is in units of bits/symbol. Spectral efficiency is an appropriate measure of the fundamental limits of this channel of since it combines important and practical channel performance measures of data rate and bandwidth. For a channel bandwidth of W_{ch} Hz using the effective dimension $\kappa(\Phi)$ (6.5), the maximum spectral efficiency is bounded as,

$$C_\eta(\Phi) \leq C_\eta^{up}(\Phi)$$

where

$$C_\eta^{up}(\Phi) =$$

$$\frac{2N}{\kappa(\Phi)} \log_2 \left[\left(\sqrt{\frac{\kappa(\Phi)}{2W_{ch}} \frac{P}{\sigma}} + 2\sqrt{N} \right) \frac{V(\Upsilon_1)^{1/N}(N-1)!^{1/N}}{N} \sqrt{\frac{e}{2\pi}} \right]$$

$$\text{[bits/s/Hz]}. \quad (6.12)$$

In the bound above, the effective dimension of each signal in Υ must be computed in order to find $\kappa(\Phi)$. Notice that this is different than in bandlimited channels where the dimension of each basis signal is unity. The factor $N/\kappa(\Phi)$ can be interpreted as a measure of *dimensional efficiency* of a given model since N is the dimension of the signal space while $\kappa(\Phi)$ is the maximum effective dimension of the set of signals determined by Φ using a $(1-\epsilon)$-fractional energy bandwidth measure.

The maximum spectral efficiency, $C_\eta(\Phi)$, does not depend on T unlike $C_s(\Phi)$. For example, signalling techniques using time-disjoint rectangular pulses is a subset of the Cartesian product of one dimensional rectangular PAM. That is, PPM can be considered as a coded version of rectangular PAM. For a given T, consider forming another rectangular pulse signalling scheme by transmitting L time disjoint pulses per period each with extent T/L seconds. As pulse width goes to zero for a fixed average optical power, $C_s(\Phi)$ can be shown to be unbounded while $C_\eta(\Phi)$ in (6.12) which is unaffected by the time extent of each of the constituent pulses. By imposing a bandwidth constraint a consistent measure of the maximum data rate of the channel is obtained independent of the particular T chosen.

6.5　Lower bound on Channel Capacity

It is possible to find a lower bound on the channel capacity of the optical intensity channel by computing the mutual information between input and output for any source distribution satisfying the amplitude constraints of the channel.

In other word, the capacity is at least as good as the mutual information between X and Y in (6.4) with source distribution $f_X(x) \in \mathcal{F}$. Consider defining a lower bound by selecting the *maxentropic source distribution*, $f_X^*(x) \in \mathcal{F}$. The distribution $f_X^*(x)$ maximizes the differential entropy of the source subject to both the non-negativity and average optical amplitude restrictions.

For a fixed T, the average optical power depends solely on the mean of the ϕ_1 coordinate value, as shown in (4.3). By the maximum entropy principle, the maxentropic source distribution subject to this average constraint must take the form $f_X^*(x) = K \exp(-\lambda x_1)$, for $x \in \Upsilon$ and for some $K, \lambda > 0$ [128]. The constants K and λ can be found by solving

$$\int_{x \in \Upsilon} f_X^*(x)dx = 1$$

$$\int_{x \in \Upsilon} x_1 f_X^*(x)dx = \sqrt{T}A$$

to yield

$$f_X^*(x) = \left(\frac{N}{\sqrt{T}P}\right)^N \frac{1}{V(\Upsilon_1)(N-1)!} \exp\left(-N\frac{x_1}{\sqrt{T}P}\right) \tag{6.13}$$

for $x = (x_1, x_2, \ldots, x_N) \in \Upsilon$. For X^* distributed as $f_X^*(x)$, the differential entropy of X^* is

$$
\begin{aligned}
h(X^*) &= -\int_{x \in \Upsilon} f_X^*(x) \log_2 f_X^*(x)dx \\
&= N \log_2 \left(\sqrt{T}P \frac{V(\Upsilon_1)^{1/N}(N-1)!^{1/N}e}{N}\right). \tag{6.14}
\end{aligned}
$$

Notice that $f_X^*(x)$ is a function of solely average optical power constraint, that is, the coordinate in the ϕ_1 direction. Conditioned on a given value of $x_1 = k$, the distribution is uniform over cross-section Υ_k, which is entropy-maximizing in the absence of constraints.

Therefore, a lower bound on the channel capacity can be written as

$$C_s(\Phi) \geq I(X^*; Y) \tag{6.15}$$

where $X^* \sim f_X^*(x)$. A closed form solution on this lower bound, however, is in general difficult to derive. A closed form on a looser lower bound on the channel capacity can be found by expanding the mutual information as

$$
\begin{aligned}
I(X^*; Y) &= h(Y) - h(Z) \\
&= h(X^* + Z) - h(Z) \tag{6.16}
\end{aligned}
$$

where $h(\cdot)$ evaluates to the differential entropy. A lower bound on $I(\boldsymbol{X}^*; \boldsymbol{Y})$ arises by

$$
\begin{aligned}
I(\boldsymbol{X}^*; \boldsymbol{Y}) &= h(\boldsymbol{X}^* + \boldsymbol{Z}) - h(\boldsymbol{Z}) \\
&\geq h(\boldsymbol{X}^* + \boldsymbol{Z} | \boldsymbol{Z}) - h(\boldsymbol{Z}) \quad (6.17) \\
&= h(\boldsymbol{X}^*) - h(\boldsymbol{Z}) \quad (6.18)
\end{aligned}
$$

where (6.17) arises since conditioning reduces differential entropy and (6.18) since translation does not alter the differential entropy.

Combining (6.14), (6.15) and (6.18) and applying the effective dimension spectral constraint (6.5) gives a lower bound on the spectral efficiency,

$$
C_\eta(\Phi) \geq C_\eta^{\text{low}}(\Phi)
$$

where

$$
C_\eta^{\text{low}}(\Phi) =
$$

$$
\frac{2N}{\kappa(\Phi)} \log_2 \left(\sqrt{\frac{\kappa(\Phi)}{2W_{\text{ch}}}} \frac{P}{\sigma} \frac{V(\Upsilon_1)^{1/N}(N-1)!^{1/N}}{N} \sqrt{\frac{e}{2\pi}} \right)
$$

$$
[\text{bits/s/Hz}]. \quad (6.19)
$$

The asymptotic behavior of the bounds (6.12) and (6.19) can be investigated at high optical SNR, when P/σ becomes large. For a fixed value of average optical power, P, the optical SNR can be increased arbitrarily by letting $\sigma \to 0$, in which case it can be shown that

$$
\lim_{\sigma \to 0} C_\eta^{\text{up}}(\Phi) - C_\eta^{\text{low}}(\Phi) = 0,
$$

that is, the bounds are asymptotically exact as optical SNR tends to infinity. Note that $C_\eta^{\text{low}}(\Phi)$ is only tight at high optical SNRs and that numerical computation of $I(\boldsymbol{X}^*; \boldsymbol{Y})$ provides a better lower bound at low optical SNR.

The relationship between $C_\eta(\Phi)$ and the bounds derived is summarized as,

$$
C_\eta^{\text{up}}(\Phi) \geq C_\eta(\Phi) \geq \frac{2}{\kappa(\Phi)} I(\boldsymbol{X}^*; \boldsymbol{Y}) \geq C_\eta^{\text{low}}(\Phi). \quad (6.20)
$$

6.6 Examples and Discussion

6.6.1 Rectangular PAM

As described in Section 3.3.2, rectangular M-PAM pulses are formed by scaling a rectangular pulse shape (4.1) so that $\Phi_{\text{PAM}} = \{\phi_1(t)\}$. The effective dimension using a 0.99-fractional energy bandwidth ($\epsilon = 10^{-2}$) is $\kappa(\Phi_{\text{PAM}}) = 20.572$, and is computed by numerical integration of the basis

Table 6.1. Comparison of main characteristics of each optical intensity model.

	Rectangular PAM	Raised-QAM	2-PSWF	3-PSWF
N	1	3	2	3
$\kappa(\Phi)$	20.572	27.038	20.572	20.572
$V(\Upsilon_1)$	1	$\pi/2 \approx 1.571$	1.914	2.173
$N/\kappa(\Phi)$ (%)	4.86	11.10	9.72	14.58

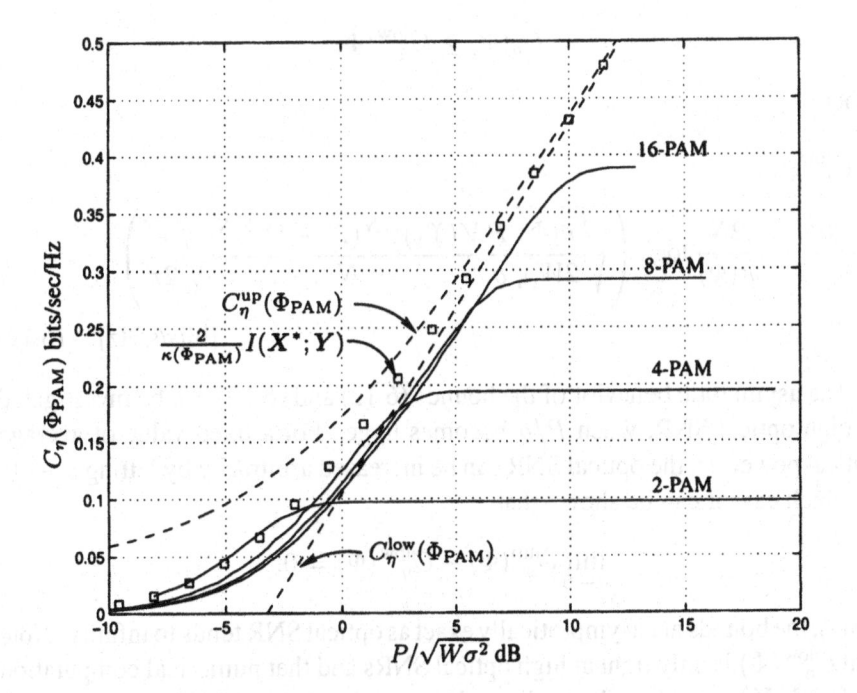

Figure 6.2. Capacity bounds and mutual information curves for rectangular PAM schemes.

function in frequency domain [123]. The cross-section Υ_1 in this case is a point and the volume is taken as 1 which allows all previous derivations to hold, and is presented in Table 6.1.

The upper and lower bounds on $C_\eta(\Phi_{PAM})$ are presented in Figure 6.2 as well as spectral efficiency curves for discrete uniform 2, 4, 8 and 16 point constellations versus optical SNR. The spectral efficiency curves for the uniformly distributed examples were computed by a well known numerical technique by Ungerboeck to compute the mutual information [142].

The upper bound on capacity (6.12) is,

$$C_\eta(\Phi_{\text{PAM}}) \leq \frac{2}{\kappa(\Phi_{\text{PAM}})} \log_2 \left[\left(\sqrt{\frac{\kappa(\Phi_{\text{PAM}})}{2W}} \frac{P}{\sigma} + 2 \right) \sqrt{\frac{e}{2\pi}} \right].$$

In order to determine the lower bounds, the received distribution, $f_Y(y)$, can be computed as,

$$\begin{aligned} f_Y(y) &= f_X^*(y) * f_Z(y) \\ &= \frac{1}{\sqrt{TP}} \left(1 - Q\left(\frac{y}{\sigma} - \frac{\sigma}{\sqrt{TP}} \right) \right) \exp\left(\frac{\sigma^2 - 2y}{2\sqrt{TP}} \right), \end{aligned}$$

where $Q(x)$ is defined in (3.8). Since $f_Y(y)$ does not have a closed form, computing $I(X^*, Y)$ in closed form is not possible. The lower bound in Figure 6.2 was computed numerically at a number of optical SNR values [123]. As shown earlier, at high SNR the lower and upper bounds approach each other.

The bounds are illustrated again in Figure 6.3 at low SNR. Spectral efficiency curves for popular M-PPM, discussed in Section 3.3.1, are also included for comparison. The spectral efficiency curves for a channel model based on M-PPM are computed via Monte Carlo simulation [142] and plotted for $N = 2, 4, 8$ in Figure 6.3. It is interesting to note that PPM schemes approach the lower bound for capacity at low optical SNR. These results also mirror previous results on PPM channels which demonstrated that higher cardinality PPM constellations asymptotically have better performance than lower size constellations as SNR decreases [113].

6.6.2 Raised-QAM

Upper and lower bounds on the capacity of raised-QAM, defined in Section 4.2.1, are plotted in Figure 6.4. The upper bound (6.12) for raised-QAM is,

$$C_\eta(\Phi_{\text{QAM}}) \leq \frac{6}{\kappa(\Phi_{\text{QAM}})} \log_2 \left[\left(\sqrt{\frac{\kappa(\Phi_{\text{QAM}})}{2W}} \frac{P}{\sigma} + 2\sqrt{3} \right) \sqrt{\frac{e}{18\pi^{1/3}}} \right].$$

The effective dimension was computed for $W_{0.01}$ using (6.6) and by noting that $|\phi_2^F(f)|^2 = |\phi_3^F(f)|^2$. The effective dimension of Φ can be shown to be $\kappa(\Phi_{\text{QAM}}) = 27.038$ and is achieved by points on the boundary of Υ. The cross-section Υ_1 is a circular disc of radius $1/\sqrt{2}$, as shown in Figure 4.7, and hence volume $V(\Upsilon_1) = \pi/2$, which is presented in Table 6.1.

As is the case with PAM, no closed from for $f_Y(y)$ exists and the lower bound was approximated using a discretized version of $f_X^*(x)$ (6.13) using a well know numerical technique [142]. As was the case for rectangular PAM, the bounds approach one another at high optical SNRs. Spectral efficiency curves

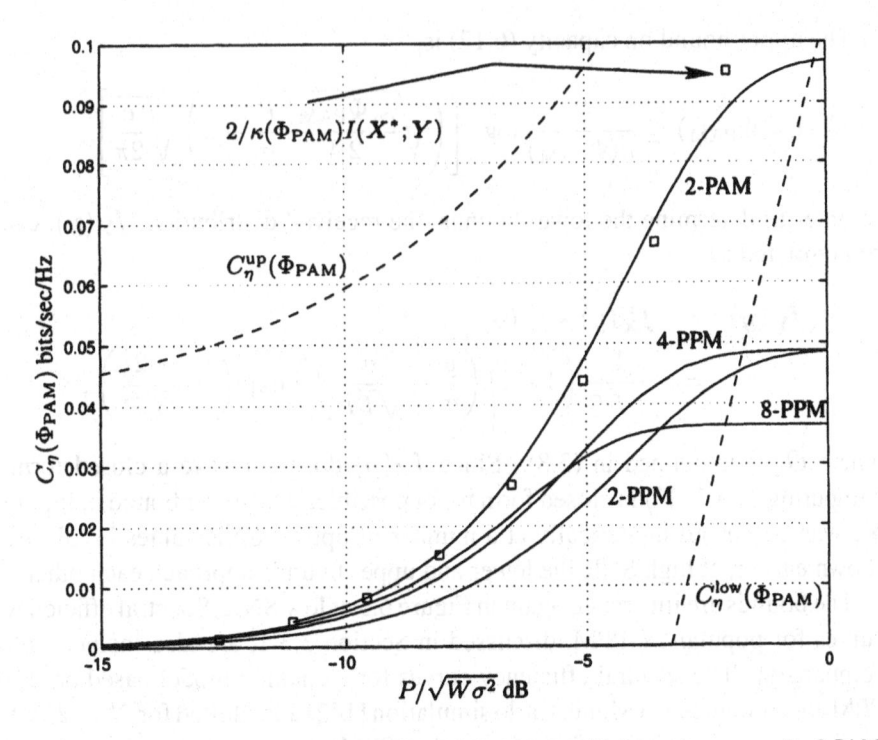

Figure 6.3. Capacity bounds for rectangular PAM with mutual information curves for 2-PAM and M-PPM schemes at low optical SNR.

for a variety of uniformly distributed raised-QAM schemes was determined using the same numerical techniques as in the PAM case [142] and are also presented in the Figure 6.4.

6.6.3   Prolate Spheroidal Wave Function Bases

Due to the use of the Landau-Pollak dimension, defined in Section 3.2, as a means to impose a bandwidth constraint, it seems natural to consider forming an optical intensity signalling scheme using the family of prolate spheroidal wave functions.

Figure 6.5(a) presents of plot of the $\varphi_0(t)$ and $\varphi_1(t)$ for $2WT = \kappa(\Phi_{PAM})$. The wave functions are approximated by highly over-sampled discrete prolate spheroidal sequences generated by a popular numerical mathematics package [143]. It can be shown that as the sampling rate goes to infinity the discrete sequence converges point-wise to the associated prolate spheroidal wave function [144].

An optical intensity signal space can be formed as described in Chapter 4 by setting $\phi_1(t)$ as in (4.1) with $2W_\epsilon(\phi_1)T = \kappa(\Phi_{PAM})$. The remaining basis functions are chosen so that $2W_\epsilon(\phi_n)T < \kappa(\Phi_{PAM})$ for $n > 1$. The

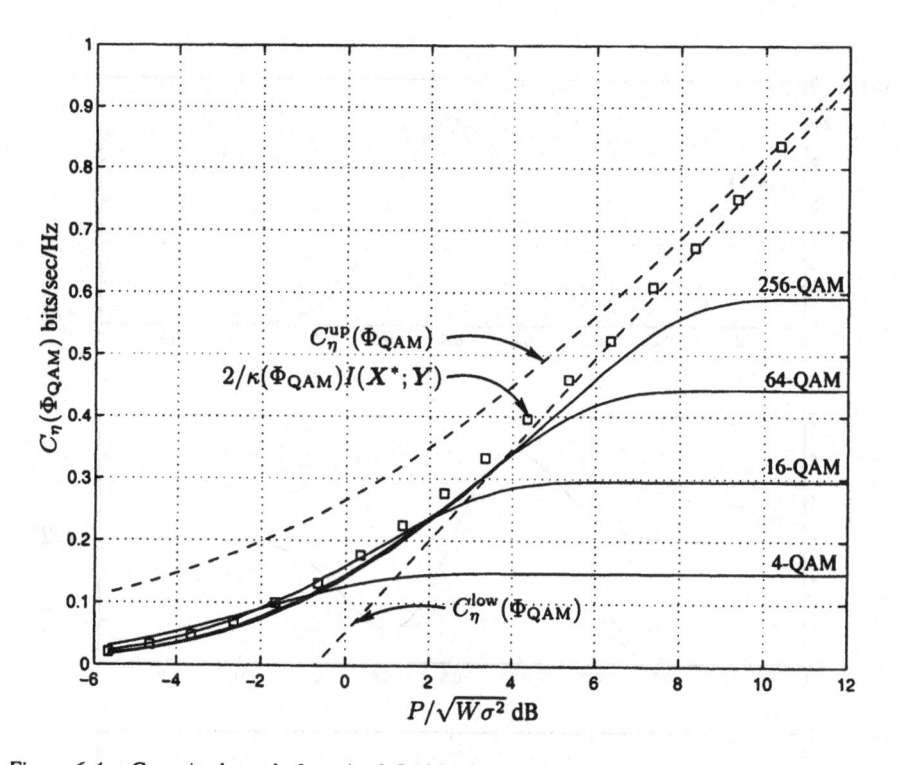

Figure 6.4. Capacity bounds for raised-QAM schemes with mutual information curves for 4, 16, 64 and 256 point constellations.

effective dimension, $\kappa(\Phi)$ can be approximated by fixing T and maximizing the bandwidth over all transmittable signals via (6.6). In this case, it can be shown that the $\phi_1(t)$ basis function has the maximum effective dimension of any unit energy signal. As a result, the effective dimension of such optical intensity sets is $\kappa(\Phi_{\text{PAM}})$.

An M-prolate spheroidal wave function (PSWF) optical intensity basis set Φ_{PSWF} is specified by performing the Gram-Schmidt orthogonalization procedure with $\phi_1(t)$ and $\varphi_m(t)$ for $m = 0, 1, \ldots, M - 2$ at a time-bandwidth product of $2W_\epsilon(\phi_1)T$. Denote these basis functions as $\phi_1(t)$ and $\varphi'_m(t)$. The cross-sectional volume, $V(\Upsilon_1)$, for 2 and 3-PSWF were computed numerically [126] and the values are given in Table 6.1. Figure 6.5(b) plots the $\text{Proj}(\Upsilon_1)$ for the 3-PSWF scheme.

Using the computed values, the upper (6.12) and lower (6.19) bounds on the spectral efficiency was computed and is presented in Figure 6.6 along with those for rectangular PAM and raised-QAM.

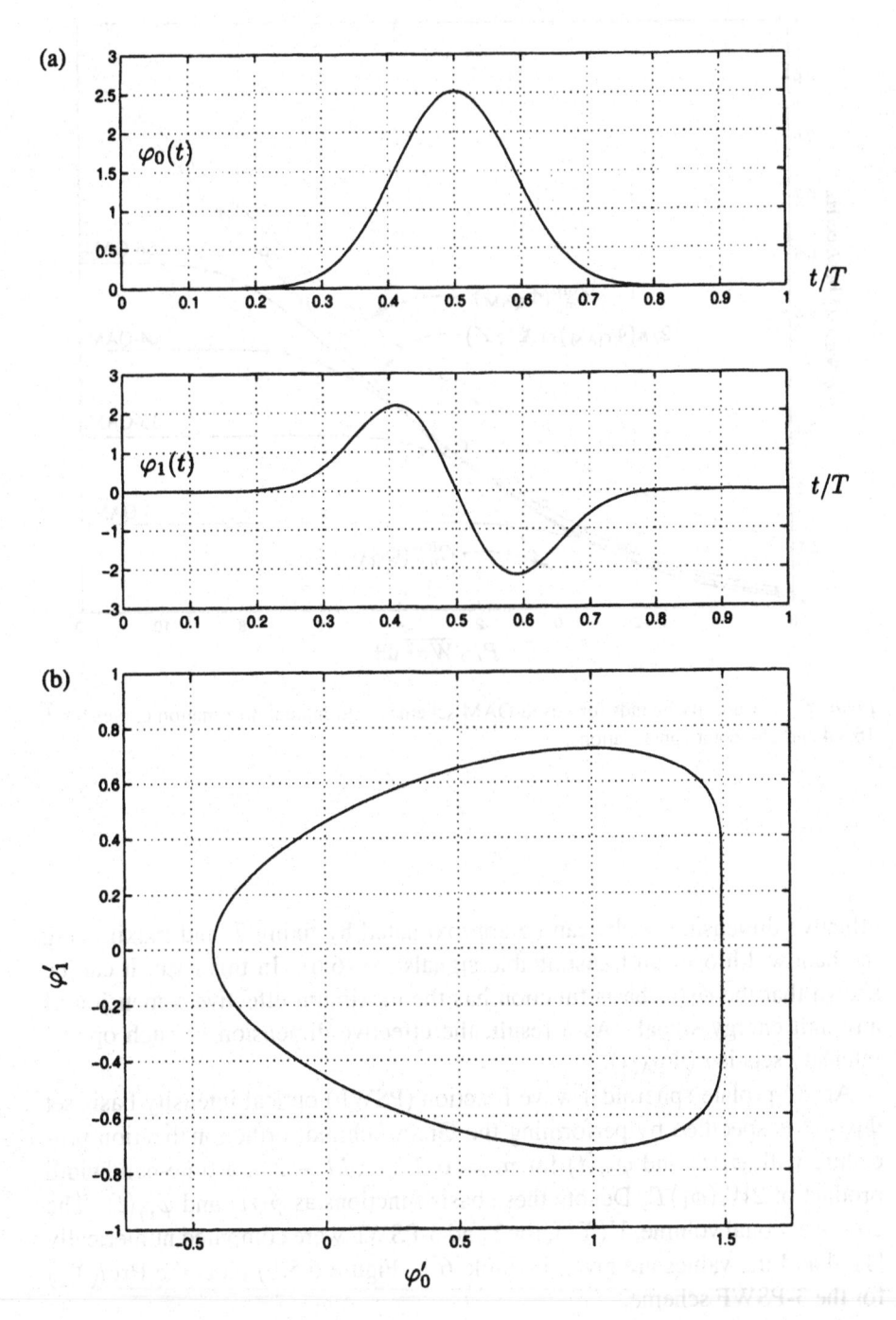

Figure 6.5. (a) Prolate spheroidal wave functions $\varphi_0(t)$ and $\varphi_1(t)$ for $\kappa(\Phi_{\text{PAM}}) = 20.572$ and (b) the corresponding $\text{Proj}(\Upsilon_1)$ region.

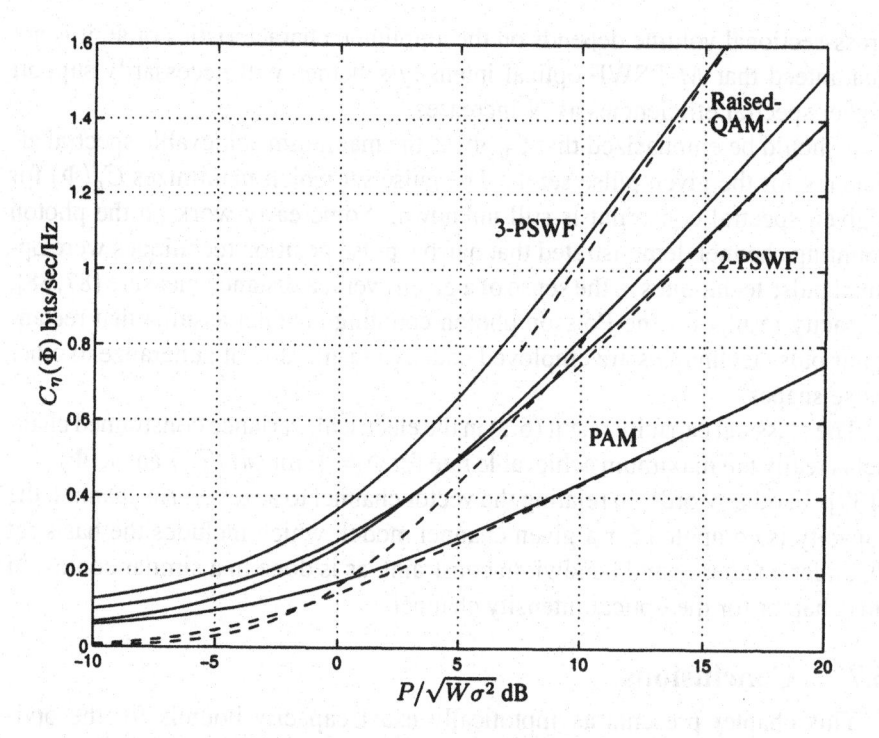

Figure 6.6. Upper and lower bounds on the spectral efficiency of rectangular PAM, raised-QAM and upper bounds for 2 and 3-PSWF (solid and dashed lines are upper and lower bounds respectively).

6.6.4 Discussion

A comparison of the capacity bounds are illustrated in Figure 6.6 for the examples considered in this chapter. At high optical SNRs, bandwidth efficient signalling schemes have a greater than twice the maximum spectral efficiencies of rectangular PAM. At low SNR, when the available spectral efficiencies are small, rectangular pulse techniques are be attractive due to their ease of implementation.

The key parameters used in the computation of the capacity bounds are presented in Table 6.1. For a given effective dimension, $\kappa(\Phi)$, the upper bound in 6.12) is maximized by the basis set Φ which simultaneously maximizes the dimensional efficiency, $N/\kappa(\Phi)$, and cross-sectional volume $V(\Upsilon_1)$. The raised-QAM basis set realizes higher spectral efficiencies by increasing the dimension of each symbol for a modest increase in the effective dimension. In the case of 2 and 3-PSWF schemes, the effective dimension is fixed to be $\kappa(\Phi_{\text{PAM}})$, however, the additional bases improve the dimensional efficiency and thus allow for larger maximum spectral efficiencies. The impact of $V(\Upsilon_1)$ on the spectral efficiency, although logarithmic, can be significant. Since the cross-sectional volume depends on the amplitude characteristics of Φ, it is not guaranteed that M-PSWF optical intensity schemes will necessarily support higher spectral efficiencies as N increases.

It should be emphasized that $C_\eta(\Phi)$ is the maximum achievable spectral efficiency for the given pulse set Φ. The pulse set which maximizes $C_\eta(\Phi)$ for a given spectral constraint is still unknown. Some early work on the photon counting channel demonstrated that narrow pulse position techniques were optimal pulse techniques in the sense of a given average distance measure [87, 88]. Capacity results for the Poisson photon counting channel assume that rectangular pulse techniques are employed and, as a result, do not generalize to other pulse shapes.

The classical capacity result (6.2) in the electrical, variance constrained channel is really the maximum achievable rate for $\Phi = \{\text{sinc}(\pi t/T)\}$ and $\kappa(\Phi) = 1$ [139]. In other words, in relating the vector channel to a waveform channel, the capacity is computed for a given channel model, which includes the basis set Φ, under an energy and bandwidth constraint, as is done in a similar context in this chapter for the optical intensity channel.

6.7 Conclusions

This chapter presents asymptotically exact capacity bounds for the optical intensity channel with average optical power and bandwidth constraints in Gaussian noise. These results are complementary to previous work on Poisson photon counting channels, since we consider the case of high intensity optical channels in which the noise can be modelled as being Gaussian and signal independent. The derived capacity bounds demonstrate that optical intensity signalling schemes based on rectangular basis sets have significantly lower achievable maximum spectral efficiencies than bandwidth efficient techniques. In particular, significant rate gains can be obtained by using raised-QAM or prolate spheroidal wave function pulse set over rectangular PAM at high optical signal-to-noise ratios.

This chapter has addressed the use of spectrally efficient modulation to achieve high dimensional efficiency, however, the use of spatial dimensions are not considered. The following chapter considers the use of spatial dimensions to provide a degree of diversity at the receiver and proposes a novel wireless optical channel architecture, the pixelated wireless optical channel, in which spatial dimensions are exploited to yield gains in spectral efficiency.

PART III

MULTI-ELEMENT TECHNIQUES

Chapter 7

THE MULTIPLE-INPUT / MULTIPLE-OUTPUT WIRELESS OPTICAL CHANNEL

The lattice codes and capacity bounds derived in Chapters 5 and 6 demonstrate the requirement of spectral efficiency in bandlimited wireless optical channels. They have shown that when the channel has a high optical SNR, the careful selection of pulse sets is a key factor in improving spectral efficiency or equivalently the dimensional efficiency. In single-transmit-element, single-receive-element systems, modulation and coding must be designed to fully exploit the number of independent dimensions available for a given bandwidth constraint.

This chapter describes the *multiple-input/multiple-output* (MIMO) wireless optical channel in which the spatial dimensions are used to improve the reliability and spectral efficiency of point-to-point links. This approach mirrors work done in radio frequency channels, except that instead of considering at most 16 transmit and receive elements, the MIMO optical channel extends the concept to an extreme dimensionality. In Chapter 8 an experimental channel is presented using on the order of 10^5 transmit elements and 10^4 receive elements to achieve large gains in spectral efficiency. Techniques developed for RF channels cannot be applied directly to the wireless optical channel due to the amplitude constraints. This chapter presents a brief survey of current multi-element techniques for communications and storage applications. The basic channel topology of point-to-point MIMO wireless optical links is presented along with a discussion on potential transmitter and receiver elements. Point-to-point wireless optical MIMO channels are discussed in greater detail and differentiated depending on the availability of spatial synchronization. Estimates on the capacity of spatially aligned, "pixel-matched", and more general pixelated wireless optical channels [145, 146] are presented.

7.1   Previous Work

The use of multiple optical transmitters to improve data throughput has been used in a number of wireless optical, chip-to-chip interconnect and holographic storage applications. The use of multiple light sources allows the transmitter to produce a number of spatially separated channels which can be used to improve channel characteristics, as in the quasi-diffuse channel of Section 2.4.3, or to improve spectral efficiency when coupled with multiple receivers, as shown in Section 7.5. Multiple receive elements afford a level of spatial diversity to the receiver. The spatial diversity allows the receiver to reject spatially localized noise sources and to separate multipath components spatially which can be used to improve channel reliability [147].

One method to realize multi-element links is to construct multiple discrete transmitters and receivers. An experimental angle diversity system was constructed with 8 transmitters and 9 receivers oriented in different directions. The resulting link operated at rates of 70 Mb/s over a 4 m horizontal range [73]. In recent work, holographic mirrors have been considered for each imager to improve the collection of optical radiation at the receiver [72]. The disadvantage of using discrete transmitters and receivers is the cost and bulk associated with each transceiver.

An alternate method to implement multi-element-receiver links is to use imaging optics. In this configuration a number of receive elements share the same imaging optics. In the original work on quasi-diffuse channels, a single fly-eye optical concentrator was used with a number of receive elements [71]. On point-to-point links, imaging receivers have been shown to have theoretical optical power gains of 13 dB over single receive element systems if the number of receive pixels is on the order of 1000 [148, 147].

Coding has also been applied to quasi-diffuse channels to improve channel reliability. A punctured Reed-Solomon code was used with PPM modulation and maximal ratio combining receivers to yield theoretical bit error rates of less that 10^{-9} at 100 Mb/s for transmitted optical powers less than a watt [149]. Punctured convolutional codes have also been applied using a code combining receiver, which performs maximum ratio combining followed by maximum likelihood sequence detection, to yield similar results [150]. It should be noted, however, that in both cases the coding is performed in time only and identical data is transmitted in all spatial directions.

Space division multiplexing, described in Section 2.4.1, employs multiple narrow beams to transmit data at high rates to various points in a room. This system can be thought of as a MIMO optical channel in which different information is transmitted in different directions. In this case, a gain the aggregate spectral efficiency is gained at the expense of requiring acquisition and tracking mechanisms.

Multi-element links have also been proposed for long range wireless optical channels. These channels differ significantly from the short-range links considered here in that they are subject to fading due to fluctuations in the refractive index of air. This fading is particularly severe at ranges of over 1 km. Space-time codes have been considered for this channel in the case of heterodyne optical detection where the phase and amplitude of the optical carrier can be measured [151, 152]. These techniques are not directly applicable to inexpensive short-range links since the optoelectronics permit only the modulation and detection of the intensity envelope. Multiple receive elements have also been considered to mitigate the impact of fading in long-range links [153]. These techniques, however, are not designed for short-range links, which are free of fading, and do not provide any gains in spectral efficiency.

Optical chip-to-chip and optical backplane applications are an instance of multi-element wireless optical links operating over distances of at most 1 m. Two-dimensional interconnects have been constructed with 256 vertical cavity surface emitting lasers (VCSELs) and 256 photodiodes, although only two links could be used simultaneously at 50 Mb/s due to technical limitations [154]. A 512 element link was also recently reported to illustrate optical design and packaging issues [155]. A common problem in all interconnects of this type is the alignment between transmitter and receiver since each pixel of the receiver is assumed to carry independent data from the transmitter [156].

Error control coding has been applied to these links with success. Reed-Solomon codes have been proposed to code each pixel in a two-dimensional array and are shown to reduce the sensitivity of the link to misalignment and improve the channel data rate and spatial density [157]. The use of lower complexity, multidimensional parity check codes has also been proposed for this channel with a resulting theoretical improvement in bandwidth efficiency when coupled with an automatic repeat request protocol [158]. Analyses have recently been extended to incorporate the effects of varying channel quality and confirm link improvement by coding each pixel with Golay as well as BCH codes [159]. A common thread through all of these studies is that the cross-talk between receive pixels is viewed as noise independent of the received values at each pixel and is not exploited in system design.

A related optical system is page-oriented optical storage or holographic optical storage. In holographic storage systems the interference pattern between a reference beam and a spatially modulated coherent source is stored in a photorefractive crystal. When the crystal is illuminated by the same reference beam the identical interference pattern is produced which is detected by an imaging array. In the typical configuration, the Fourier transform of the binary-level spatial light modulator output is stored in the medium by using lens system. Apertures are typically used to spatially low-pass filter the images before storage, essentially accomplishing a crude form of image compression. At the receiver, the inverse Fourier transform is accomplished by a similar lens system [160, 161].

In holographic storage systems strict spatial alignment is usually present so that each receive element images a single transmit pixel [162, 163]. These "pixel-matched" systems, however, are very sensitive to misalignment errors. Additionally, small deviations in the reference beam during reconstruction can lead to errors in the output image. The primary channel impairment in volumetric holographic storage is due to cross-talk between pages as well as medium non-linearities [164]. A number of algorithms have been developed for these two-dimensional channels such as spatial equalization [165], partial response [166], interpolation [167, 168] and coding schemes [169, 170]. These algorithms, although designed for a different medium, may also be applicable to the pixelated wireless optical channel defined in Section 7.5.

7.2 The MIMO Wireless Optical Channel

7.2.1 Background

Traditional modem practice for the optical intensity channel consists of the design of time-varying signals which efficiently satisfy the channel constraints. The spatial distribution of the optical intensity at the transmitter and receiver is not exploited to improve the spectral efficiency of the link.

In traditional point-to-point links, discussed in Section 2.4.1, the spatial dimensions inherent in the channel cannot be exploited due to the directed nature of the channel. Diffuse and quasi-diffuse links, described in Sections 2.4.2 and 2.4.3, transmit the same signals to all locations and act as an inefficient repetition code in space.

The MIMO wireless optical channel optical channel is a multi-element link which exploits spatial dimensions to achieve gains in reliability and spectral efficiency. These gains are achieved by implementing a transmitter which replaces the spatial repetition code with a more efficient code. The receiver is composed of a number of receive elements which detect the radiant optical power from a number of spatial modes. The gains in spectral efficiency can be realized by considering coding in time and in space, i.e., *spatio-temporal* coding. Thus, the problem of modem design becomes one of designing a series of time-varying *images* which are detected by an array of receive elements.

The transmitter, in the most general terms, is a *spatial light modulator* which produces an output optical intensity spatial distribution which is controlled by optical or electrical addressing [171]. In other words, the transmitter outputs a time-varying optical intensity image which is transmitted in free-space. Two-dimension examples of such transmitters are liquid crystal displays and arrays of light emitting diodes. Section 7.2.2 presents some examples of possible transmitting elements.

The receiver produces an output signal representing the spatial distribution of optical power impinging on the device. In most practical implementations, this device can be thought of as a temporal and spatial optical intensity sampler. Typical two-dimensional examples of such receivers are charge-coupled device (CCD) cameras as well as arrays of photodiodes. Some examples of these receivers are presented in Section 7.2.3.

The point-to-point MIMO wireless optical channel differs from the holographic storage channel, described in Section 7.1. The MIMO wireless optical systems considered here operate under incoherent illumination and are not affected by medium non-linearities as are holographic storage systems. Additionally, the requirement for strict spatial alignment between the pixels of the transmitter and the receiver is not present in the pixelated wireless optical channel of Section 7.5. The primary noise source in MIMO wireless optical channels is due to ambient light sources and is signal dependent, as measured in Section 8.2, whereas holographic storage suffers from inter-page cross-talk. However, holographic systems, like MIMO wireless optical links, are page oriented devices which transmit and receive data in two-dimensional arrays. As a result, algorithms developed for point-to-point MIMO wireless optical channels may also be applicable to holographic storage systems.

Figure 7.1 presents an example of a point-to-point MIMO wireless optical channel. In this example, the transmitter produces a planar two-dimensional intensity wavefront from an array of light emitting diodes. This transmitter forms a series of spatially discretized images by assigning intensity values to the individual pixels of the transmit array. Each transmitted image can be viewed as a codeword taken from the spatio-temporal codebook. The transmitter is positioned so that its optical intensity field is in the field of view of the receiver. The receiver, potentially an array of photo-sensitive devices, samples the optical field over the field-of-view. The received information is processed and decoded.

The point-to-point MIMO wireless optical channel topology can also be applied in a quasi-diffuse application as shown in Figure 7.2. The transmitter forms an image on the ceiling of the room in question. The receiver is oriented so as to capture the image projected on the ceiling. In this manner, a link can be established without need for a line-of-sight between transmitter and receiver.

7.2.2 Potential Transmitters

The transmit spatial light modulator can be realized in a number of present day technologies.

Liquid crystal display (LCD) panels have been used for some time in display applications. These devices modulate light through the electrical control of the polarization of a liquid crystal sandwiched between two oppositely oriented polarizers [171]. For display applications which operate in discrete-time, 1280×1024 pixel arrays are available operating at 24 bits/pixel color depth

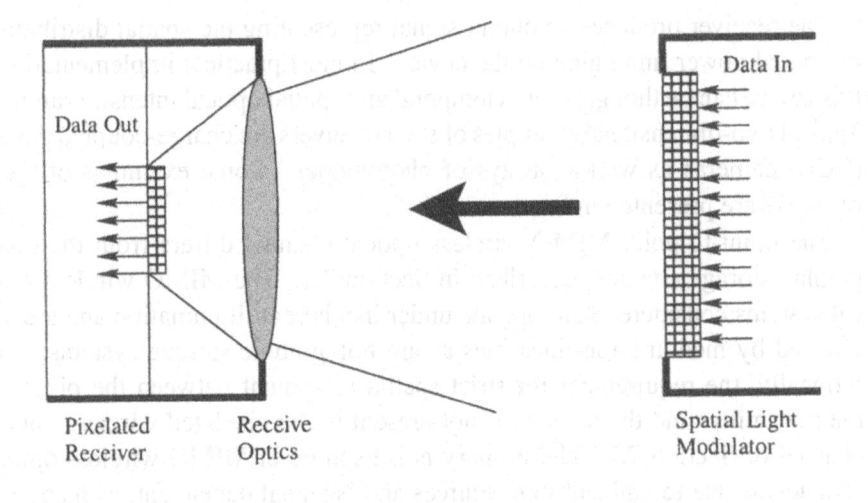

Figure 7.1. A point-to-point pixelated wireless optical channel.

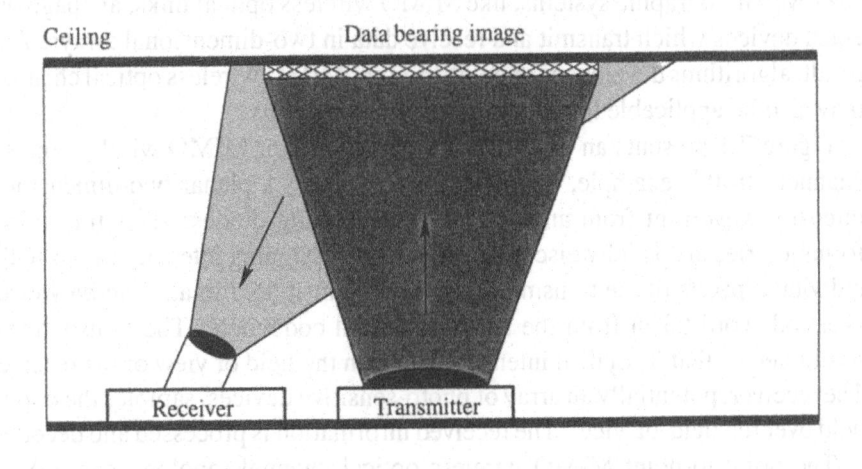

Figure 7.2. A quasi-diffuse pixelated wireless optical channel.

at a maximum frame rate of less than 100 Hz. The pixel elements can also be modulated continuously in time.

Deformable mirror devices (DMD) are a new class of spatial light modulator. These devices fall into a wider class of devices known as micro-electromechanical (MEM) devices. They consist of an array of mirrors which can be deflected electrostatically to modulate a constant light source. These discrete-time modulators can operate at switching speeds of approximately 100 kHz with array sizes of up to 2048 × 1152 pixels [172]. Typically applications include high-definition television units.

The spatial light modulator used in the definition of the MIMO wireless optical channel need not be planar. Organic polymer light emitting diodes

(OLEDs) and electronic ink (E-INK) technologies allow flexible transmitters to be realized. OLEDs are emissive transmitters printed on flexible sheets of plastic using commercial printing processes. A 15-inch prototype display was manufactured with 1280×720 pixels with a microsecond switching time [173]. E-INK is a reflective display technology which offers high output resolution at low frame rates. An E-INK display with a 160×240 pixel array and a thickness of 0.3 mm was recently fabricated. The display can be bent to a radius of curvature of 1.5 cm with no degradation in performance. Switching speed, however, is limited to 250 ms per page [174]. Future hemispherical optical transmitters formed using OLED and E-INK devices with a large number of pixels can be envisioned for communications purposes.

Wireless optical communication need not only take place at infrared wavelengths. Indeed, others have proposed using visible illumination devices in a dual role as communications devices [16, 17]. The potential exists, however, to use arrays of white illumination LEDs to construct multi-element optical wireless links.

The use of arrays of lasers for high-speed chip-to-chip communications, discussed in Section 7.1, has been considered to improve data rates [175, 176]. This multi-transmitter/receiver link can be thought of as a pixelated wireless optical channel. Potential rates are approximately 500 Mbps per pixel on an array of 3×3 pixels [175]. Arrays as large as 512 elements have been constructed to demonstrate fabrication and packaging issues [155].

7.2.3 Potential Receivers

The receiver is composed of an array of photodetective elements. The simplest implementation of such a structure can be accomplished by an array of photodiodes each with their own imaging optics [71, 73]. Although simple, the disadvantage of this approach is that relatively few receive elements can be used due to the cost and large size of the optical concentrators.

Angle diversity receivers, described in Section 7.1, integrate a planar array of photodetectors together with a common single optical imaging system. Viewed differently, these receivers can be used as the multi-element receivers required for the pixelated optical channel.

A charge-coupled device (CCD) imager is a popular architecture for imaging arrays. This sensor is divided into an array of pixels. Each pixel contains a metal-oxide-semiconductor capacitor which is biased to form a depletion region under the gate. The impinging optical intensity field photo-generates a charge in this region proportional to the number of arriving photons. After a predetermined integration time, known as the *frame interval*, the charge from all pixels is transferred serially to a small number of voltage-sense amplifiers which convert the charge in each pixel to a voltage level. This device is termed *charge-coupled* since the charge is shifted from pixel-to-pixel with the use of properly phased clock signals [177]. High quality imagers can be formed with the CCD architecture. Typical video imaging array (Sony ICX074AL) operates with approximately 330 kpixels at a frame rates in excess of 60 frames per second [178].

Imaging arrays have also been constructed using commodity CMOS silicon processing techniques. These arrays allow data to be read from the photosensing array in a random fashion, much like reading from memory. It is also possible to integrate signal processing within each of the pixel elements to improve image quality. A 64×64 pixel array operating at 2.5 Mbps was constructed incorporating transimpedance amplifier, filtering, gain control and thresholding per pixel [179]. An imaging 7-pixel receiver with integrated optics and control electronics has been constructed for a 155 Mb/s point-to-point wireless optical link [180]. Another example is a 352×288 pixel imager that was constructed with single slope 8-bit analog-to-digital conversion at each pixel while operating at 10 kframes per second [181]. The main disadvantage of these sensors is that in order to achieve a certain fill-factor these devices require large pixel sizes. These problems will decrease as CMOS technology continues to scale smaller in feature size [182].

7.2.4 Assumptions and Channel Model

A prototypical MIMO wireless optical channel is illustrated in Figure 7.3. Here it is assumed that the transmitter has n_T identically shaped transmit pixels distributed and a square $n_{Tx} \times n_{Ty}$ grid at intervals of D_T. Due to the amplitude constraints of the channel, at time instant t the transmitted amplitude must satisfy $a[m, n; t] \geq 0$. The transmitted optical intensity image at time t is defined as

$$s(x, y; t) = p_T(x, y) \otimes \sum_{m,n} a[m, n; t]\delta(x - mD_T, y - nD_T), \qquad (7.1)$$

where $p_T(x, y)$ is the optical intensity distribution associated with each transmit pixel, $\otimes$ is two-dimensional convolution and $\delta(x, y)$ is defined by

$$\int \int f(x, y)\delta(x - x_0, y - y_0)dxdy = f(x_0, y_0).$$

For simplicity, assume that the time modulation for each pixel is a rectangular PAM signal with symbol interval T, i.e., for $t \in (iT, (i+1)T]$ then $a[m, n; t] = a[m, n; iT]$. This time modulation format is compatible with the transmitters and receivers discussed in Sections 7.2.2 and 7.2.3. Additionally, it is possible to transmit multiple images in each symbol interval by multiplexing over different wavelengths. In this chapter, however, the basic case of transmission using the intensity of a single wavelength channel is considered.

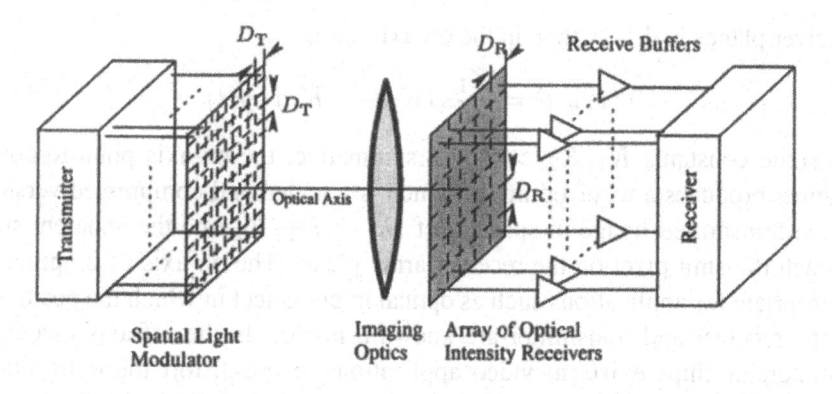

Spatial Light Imaging Array of Optical
Modulator Optics Intensity Receivers

Figure 7.3. Block Diagram of a point-to-point MIMO Wireless Optical Channel.

The receiver is positioned to collect the transmitted optical intensity image and outputs a signal representing the spatial distribution of optical power impinging on the device in each symbol interval. Denote the receive optical intensity image at time t as $r(x, y; t)$. In this channel model, the receiver consists of n_R pixels of shape $p_R(x, y)$ spaced at on a square grid of size $n_{Rx} \times n_{Ry}$ with intervals of size D_R. The receiver outputs samples, $r[k, l; iT]$, in time and in space of the optical intensity distribution. At each symbol interval, the receive array produces the sampled output of spatial intensity integrator which can be represented as,

$$r[k, l; iT] = \int_{t \in (iT, (i+1)T]} \int_{p_R(x-kD_R, y-lD_R)} r(x, y; t) d(x, y) dt + n[k, l; iT],$$

(7.2)

where $n[k, l; iT]$ is the noise process at each received pixel. It is not uncommon for these noise sources vary over the array as a result of device tolerances and non-uniform optical response of the pixels [177]. The noise in this channel is also well modelled as being Gaussian due to the intense background illumination and is characterized for an experimental channel in Section 8.2.3. Additionally, perfect temporal synchronization between the receiver and transmitter is assumed and temporal inter-symbol interference is also neglected. An optical imaging element is required to produce a focused image on the surface of the detector array.

The transmitter and receiver are at fixed positions and the channel characteristics are assumed to be static in time. This is a realistic assumption for free-space optical backplane applications or when the varies slowly and can be tracked accurately. To simplify analysis, it is assumed that the optical axes of the transmitter and the receiver are aligned. This on-axis configuration produces a received image which is an *orthographic projection* of the transmitted image free of perspective distortion [183]. If the distance between the transmitter and receiver planes is $d > 0$, then in the on-axis case,

$$r(x, y; t) = \frac{K_1}{d^2} s\left(K_2 d \cdot x, K_2 d \cdot y; t\right) \tag{7.3}$$

for some constants $K_1, K_2 > 0$. To summarize, the on-axis point-to-point channel produces a received image which is a scaled and compressed version of the transmitted image in space. Let $D'_T = D_T / K_2 d$ be the apparent size of each transmit pixel on the receiver array plane. The on-axis assumption is appropriate for applications such as optical interconnect in which the positions of the receiver and transmitter are known *a priori*. In the off-axis imaging, commercial chips exists in video applications to pre-distort the transmitted image to minimize the projective distortion in the projected image [184].

A spatially invariant channel is characterized by a *point-spread function*, which is the response of the channel to a spatial impulse [185]. Section 8.2.1 presents measurements on a experimental channel which demonstrate that the point-spread function is approximately spatially invariant and low-pass in spatial frequency domain. Let $h(x, y)$ denote the point-spread function from the transmitter to the receiver plane to account for linear systems given in (7.1), (7.2) and (7.3) to yield,

$$r[k, l; iT] =$$

$$h(x, y) \otimes \sum_{m,n} a[m, n; iT] \delta(x - m D'_T, y - n D'_T) \Bigg|_{(x,y)=(k \cdot D_R, l \cdot D_R)} + n[k, l; iT],$$

which simplifies to

$$r[k, l; iT] = \sum_{m,n} a[m, n; iT] h(k D_R - m D'_T, l D_R - n D'_T) + n[k, l; iT]. \tag{7.4}$$

The point-spread function, $h(x, y)$, will also include the impact due to lens abberations and diffraction effects in addition to the factors mentioned here [185].

7.3   Design Challenges

This section presents the primary challenges that a modem for the MIMO wireless optical channel must overcome. These challenges arise due to the channel impairments, discussed in Section 7.2.4, as well as due to practical issues such as temporal and spatial synchronization. Chapter 8 presents a channel model which measures these impairments in an experimental channel to construct a simulation model.

7.3.1   Spatial Frequency Response

For spatially invariant systems, the point-spread function is independent of the location of the intensity point source. The Fourier transform of the point-spread function is called the *optical transfer function* and determines the spatial frequency response of the system [185, 186].

The spatial frequency response of any optical system is fundamentally limited by diffraction. Diffraction is the tendency of light to bend when passing through opaque apertures or near sharp edges. Other non-idealities also limit the spatial frequency response. Aberrations in the lens optics and focusing errors increase the size of the point-spread function [185]. Additionally, the spatial averaging of the CCD array will limit the spatial frequency response which will result in an overall low-pass response.

The channel spatial frequency response, measured in Section 8.2.1, imposes a limit on the maximum rate at which the transmitted spatial image can vary. The channel response introduces spatial inter-symbol interference (ISI) in the received image which appears as a blurring or smearing of the image with respect to the transmitted image.

A more severe form of distortion arises due to aliasing distortion at the receiver. If the input signal from the channel is not bandlimited, the sampled output of the CCD sensor will be corrupted by spatial frequency aliasing distortion. In general, to guarantee that the input to the CCD sampler is spatially bandlimited, a spatial low-pass filter should be applied to the input. An incoherent spatial low-pass filter can be implemented using a transparency applied before the receive lens, using focusing error or in the frequency domain [185].

7.3.2 Channel Distortion

Cathode ray tubes and LCD panels have a non-linear relationship between the input level to be transmitted and the optical intensity output. This non-linearity is commonly referred to as *gamma distortion* [177]. Typical commercial video cameras have gamma correction circuitry to linearize the optical intensity response of the channel. The channel can be linearized by predistorting the transmit image with the inverse non-linearity, assuming that the transmitter has knowledge of this function. Section 8.2 measures this non-linear distortion using commercial video equipment.

Noise in CCD imagers is due to a variety of sources including shot noise and amplifier noise. These noise sources are also non-uniform in space as a result of electronics tolerances and non-uniform photo-response of the array [177]. These noise sources manifest themselves as fluctuations in the grey-level value that is output from the sensor. The field-of-view of each pixel is small and, as mentioned in Section 2.3, the assumption of signal independence due to the background illumination is no longer valid. The distribution of this noise is typically taken as being Gaussian and signal dependent, and is measured in Section 8.2. In addition to this noise, quantization noise is also present due to the receive quantizer.

In this channel the primary constraint on the transmitted optical power arises due to the peak limitation of the transmitter. This is in contrast to the wireless optical channel treated earlier. This peak restriction arises due to the components selected to form the link.

7.3.3   Spatial Registration and Synchronization

In order to be able to detect the incoming images, the receiver must locate the transmitted in the field-of-view. The process of determining the location of an object in the field-of-view of a camera with respect to the camera coordinate system is known as *registration*. Image registration is typically done in machine vision applications during camera calibration [183]. Registration is typically done with the aid of special calibration symbols, known as , which are known *a priori* at the receiver. Fiducial markings are typically added to printed circuit boards to aid in component placement and in very-large scale integrated circuit manufacture to align masks. The design of fiducials which allow for sub-pixel accuracy has been investigated [187, 188].

It must be noted, however, that although the receiver can infer the position and orientation of the transmitter from the captured images, it is not possible to adjust the phase or frequency of the spatial sampling. In other words, spatial synchronization is not possible since the sampling phase and frequency are determined by the pixel size of the CCD and the separation of transmitter and receiver.

7.3.4   Temporal Synchronization

In order to have a functional link, the receiver must be synchronized to the transmitter in time as well as in space. A simple technique to accomplish temporal synchronization is to transmit the frame clock as part of the transmitted image. Although this technique is simple to implement, it is an inefficient use of the transmit pixel. Well known inductive and deductive timing recovery techniques can be applied to the received data signal to extract the clock at the expense of greater complexity [45].

7.4   Pixel-Matched System

The simplest MIMO wireless optical system is one in which each transmit element corresponds to a unique receive pixel. This channel is termed *pixel-matched* and consists of a series of parallel and independent channels. Consider the case when $n = n_T = n_R$, $D'_T = D_R$ and

$$h(x, y) = \begin{cases} 1 & \text{if } x = y = 0, \\ 0 & \text{otherwise.} \end{cases}$$

The channel model (7.4) simplifies to

$$r[k, l; iT] = a[k, l; iT] + n[k, l; iT],$$

and the MIMO optical channel is a collection of n sub-channels with no inter-channel interference. For the sake of computation, let the noise process $n[k, l; iT]$ be independent identically distributed Gaussian random variables.

Let each pixel in the transmitted be modulated with an M-PAM signalling scheme, as defined in Section 3.3.2, independent of all other transmit pixels. Additionally assume that the average optical power limit is at most P and that, due to symmetry, each sub-channel is allocated an average optical power limit of $P_i = P/n$. If the size of the array is fixed, increasing n implies that the area of each receive element decreases as n. If the illumination is intense then the channel is operating shot-noise limited mode and the variance of the photo-generated noise in each sub-channel is proportional to the area of the pixel, i.e., $\sigma_i^2 = \sigma^2/n$, where σ^2 is the variance of the sum over all noise components pixels [2].

As discussed in Chapter 6, the capacity of each sub-channel is not known in closed form. The capacity of each sub-channel was estimated in this case by an electrical channel of the same variance as the optical intensity constellation. The signal-to-noise ratio in each M-PAM sub-channel is related to the average optical power as,

$$\text{SNR}_i = \frac{M^2 - 1}{3(M - 1)^2} \frac{\kappa P^2}{2W\sigma^2} \frac{1}{n} \tag{7.5}$$

where W and κ are defined in Chapter 5. The capacity of the pixel matched system can be approximated as the sum of the capacities of the individual sub-channels as,

$$C \approx \frac{n}{\kappa} \log \left(1 + \frac{\kappa(M^2 - 1)}{6(M - 1)^2} \frac{P^2}{W\sigma^2} \frac{1}{n} \right) \quad \text{[bits/s/Hz]} \tag{7.6}$$

As $n \to \infty$,

$$C \to \frac{M^2 - 1}{6(M - 1)^2} \frac{P^2}{W\sigma^2} \frac{1}{\ln 2} = \frac{M^2 - 1}{6\ln(2)(M - 1)^2} \text{OSNR}^2,$$

where OSNR is defined as $P/\sqrt{W\sigma^2}$. Therefore, as the number of pixels increases the capacity grows *quadratically* with the OSNR. Figure 7.4 plots the capacity estimate in (7.6) for $M = 2$ and OSNR=10 dB along with the results by numerically computing the mutual information of the optical constellation for uniform signalling. The figure indicates that a significant gain in spectral efficiency can be realized by using multiple pixels.

This conclusion mirrors a similar result in fading-free radio channels in which spatial multiplexing gains are realized by distributing the power constraint over a large number of spatial degrees of freedom [189, 190].

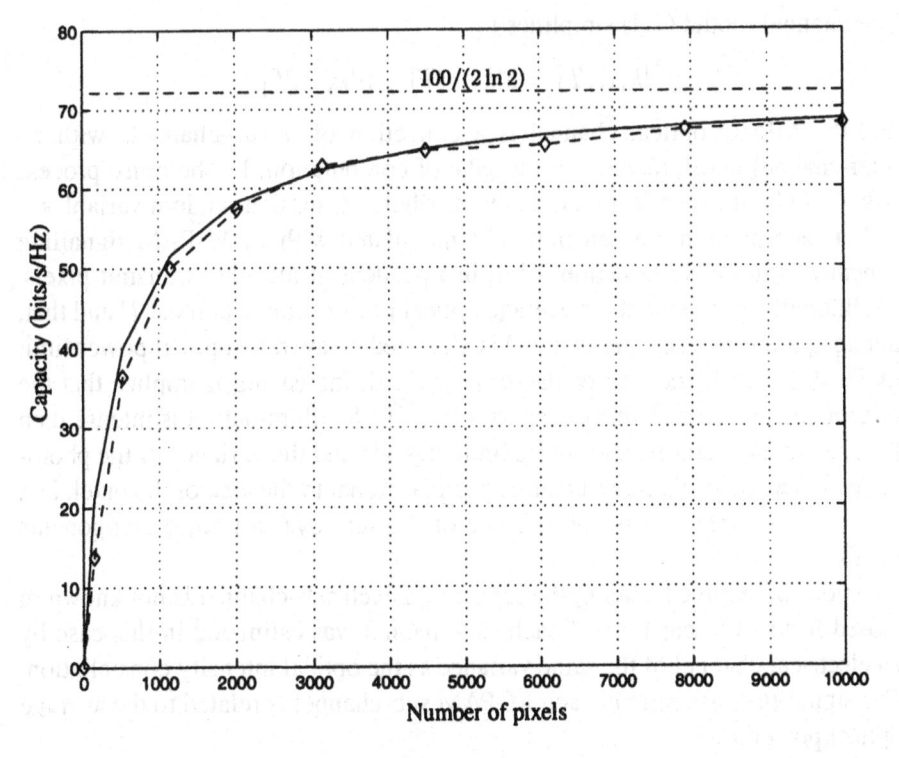

Figure 7.4. Capacity (in bits/s/Hz) as a function of number of pixels for OSNR=10 dB, 2-PAM pixel-matched system. (solid line is (7.6) points indicated by $\lozenge$ are generated by numerically computing the mutual information).

The pixel-matched channel can realize significant gains in spectral efficiency over single element case by using the large number of degrees of freedom available. These gains in spectral efficiency, however, are only significant when the OSNR is significantly large.

The OSNR in typical systems is limited due to safety consideration, however, short-range links have less losses than High-OSNR channels occur in short-range wireless optical links which limit geometric losses. Typical examples of short-range links include board-to-board and optical backplane applications and interconnection applications. Alternatively, the eye safety limit is significantly relaxed at wavelengths over 1.3 μm range allowing an increase in the transmitted optical power over conventional devices, as mentioned in Section 2.1.2.

7.5   The Pixelated Wireless Optical Channel

The pixel-matched MIMO wireless optical channel provides gains in spectral efficiency at the expense of requiring exact alignment between the transmit and receive elements. Furthermore, practical pixel-matched systems always suffer from some inter-channel interference and the number of pixels that can be used

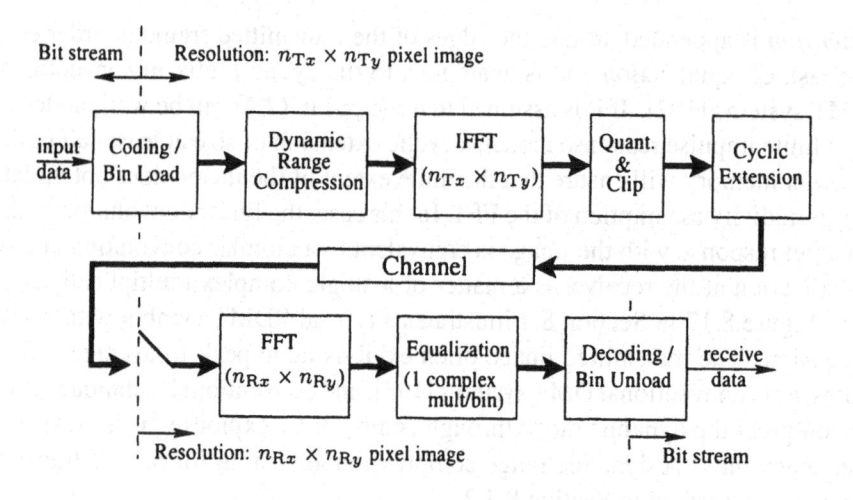

Figure 7.5. Block diagram of SDMT system.

is typically limited by this interference. In this section we consider the more general case of the point-to-point *pixelated wireless optical channel* [145, 146] in which the spatial response of the channel is taken into account and spatial alignment is not required.

7.5.1   Spatial Discrete Multitone Modulation

In frequency selective channels, discrete multitone (DMT) modulation is a popular signalling scheme, especially for digital subscriber lines [191–193]. In this modulation scheme, the frequency spectrum is divided into a number of non-overlapping bins. These bins can be viewed as a set of parallel Gaussian channels. Subject to a total electrical power constraint, the optimal power allocation across the bins is given by a "water-pouring" spectrum [128]. In electrical systems, QAM constellations are transmitted in each frequency bin with power in each bin dictated by the optimal power allocation.

Define *spatial discrete multitone* (SDMT) modulation as an extension to the DMT in which data is transmitted by modulating in the spatial frequency domain. A block diagram of the SDMT system is presented in Figure 7.5. The data is encoded and loaded into spatial frequency bins depending on the SNR in each sub-channel. A training period must preceding data transmission to characterize the channel and to allow the receiver to feed back channel information to the transmitter. The transmitter computes a power allocation based on the channel spatial characteristics and loads the spatial frequency bins appropriately. In Section 8.4 a power allocation is presented for an experimental channel.

The transmitted image is formed by performing the inverse fast Fourier transform (IFFT) of the $n_{Tx} \times n_{Ty}$ image in spatial frequency domain. A *cyclic extension* is appended around the edges of the transmitted frame in order ease the task of equalization and is analogous to the cyclic prefix in conventional DMT systems [192]. If it is assumed that $h(x, y)$ in (7.4) can be well modelled by a finite impulse response model, a cyclic extension of size at least half of the channel memory will ensure that the finite extent of the image does not violate the periodicity assumption of the FFT. In this case, the linear convolution of the channel response with the image is equivalent to a circular convolution and so equalization at the receiver is a matter of a single complex multiplication per bin. Figure 8.17 in Section 8.4 illustrates a typical SDMT symbol with cyclic extension. The transmitted image often exhibits large peak-to-average amplitudes, as in conventional DMT systems [193], and conventional techniques exist to compress the dynamic range through coding or by exploiting unused spatial frequency bins. A dynamic range compression algorithm for SDMT transmit images is described in Section 8.4.2.

At the receiver, the imaging array samples the incoming intensity signal in space. Although the transmitted signal is designed to be bandlimited to the Nyquist region [194] of the spatial sampler, aliasing is an impairment at the front end of the receiver due to the clipping and quantization noise generated in the transmitter. A $n_{Rx} \times n_{Ry}$ point fast Fourier transform (FFT) is used to place the received image into frequency domain. The resulting frequency bins are equalized, decoded and unloaded. Thus, SDMT modulation provides a basis for the set of signals spatially bandlimited to the Nyquist region of the imaging array and tailors the transmitted image to minimize the impact of the spatial distortion of the channel.

7.5.2   Spatial Synchronization for SDMT

The spacing of the receive imager specifies the spatial bandwidth region of signals detected by the receiver. The transmitted image, $s(x, y; iT)$, is designed so that the data lies within the Nyquist region of the receiver. Therefore, in a non-distorting channel it possible to recover the transmitted data from the received signal, $r[k, l; iT]$. In conventional bandlimited channels, it is not necessary to have the frequency or phase of the receiver sampler synchronized to that of the transmitter, a free-running oscillator operating above the Nyquist rate is enough to ensure no information is lost [195]. In this section, this result it explored in the case of SDMT to demonstrate that strict alignment of transmit and receive pixels is not required to eliminate inter-channel interference.

Consider a $n_{Tx} \times n_{Ty}$ pixel image with complex frequency domain coefficients $A[u, v]$ and let $H[u, v]$ be the sampled Fourier spectrum of the point-spread function $h(x, y)$. Assuming the use of a cyclic extension, the resulting received image can be modelled as the periodic image,

$$r(x, y) = \frac{1}{n_T} \sum_{u,v} H[u, v] A[u, v] \exp\left(2\pi j \frac{1}{D_T'} \left(\frac{ux}{n_{Tx}} + \frac{vy}{n_{Ty}}\right)\right).$$

Since the impact of receive pixel shape is considered in $h(x, y)$, the receive imager is modelled as a rectangular "bed-of-nails" spatial sampler. The sampled received signal takes then takes the form,

$$r[k, l] = \frac{1}{n_T} \sum_{u,v} H[u, v] A[u, v] \exp\left(2\pi j \frac{D_R}{D_T'} \left(\frac{uk}{n_{Tx}} + \frac{vl}{n_{Ty}}\right)\right).$$

Taking the $n_{Rx} \times n_{Ry}$ point FFT of $r[k, l; iT]$ yields the signal $R[w, z; iT]$ in frequency domain,

$$R[w, z] = \frac{1}{n_T} \sum_{k,l} \sum_{u,v} H[u, v] A[u, v]$$

$$\exp\left(2\pi j \frac{D_R}{D_T'} \left(\frac{uk}{n_{Tx}} + \frac{vl}{n_{Ty}}\right)\right) \exp\left(-2\pi j \left(\frac{kw}{n_{Rx}} + \frac{lz}{n_{Ry}}\right)\right).$$

Expanding the sum over k, l,

$$R[w, z] = \frac{1}{n_T} \sum_{u,v} H[u, v] A[u, v]$$

$$\sum_{k=0}^{n_{Rx}-1} W_{n_{Rx}}^{\frac{n_{Rx} D_R}{n_{Tx} D_T'} uk} W_{n_{Rx}}^{-kw} \sum_{l=0}^{n_{Ry}-1} W_{n_{Ry}}^{\frac{n_{Ry} D_R}{n_{Ty} D_T'} vl} W_{n_{Ry}}^{-lz} \quad (7.7)$$

where $W_M = \exp(j2\pi/M)$. Notice that the last two terms in (7.7) can be interpreted as n_{Rx}- and n_{Ry}-point discrete Fourier transforms respectively. If the conditions,

$$n_{Rx} D_R = n_{Tx} D_T' \quad \text{and} \quad n_{Ry} D_R = n_{Ty} D_T', \quad (7.8)$$

then

$$R[w, z] = \frac{n_R}{n_T} \sum_{u,v=0}^{N-1} H[u, v] A[u, v]$$

$$\delta[(u - w) \bmod n_{Rx}] \, \delta[(v - z) \bmod n_{Ry}].$$

Since the transmit signal is bandlimited to the Nyquist region of the receive spatial sampler, i.e., $A[u, v] = 0$ for

$$u \notin \{0, 1, \ldots, n_{Rx} - 1\} \quad \text{and} \quad v \notin \{0, 1, \ldots, n_{Ry} - 1\},$$

then

$$R[w, z] = \frac{n_{\text{R}}}{n_{\text{T}}} A[w, z].$$

The conditions in (7.8) ensure that the frequency resolutions of the transmitter rand receiver are identical are identical, and if they are not satisfied inter-channel interference (ICI) is unavoidable. The spatial synchronization problem now reduces to one of ensuring conditions (7.8) are met, i.e., that the dimensions of the projected transmit image are an integer number of receive imager sampling instants. In the case of transmitting a square array, $n_{\text{T}x} = n_{\text{T}y}$, the conditions (7.8) are equivalent. The scale of the transmitted image in the receive plan can be adjusted to satisfy (7.8) if a multi-focal length lens is available at the receiver camera. Alternatively, if the data is restricted to a bandwidth region contained inside the Nyquist region, it is possible to interpolate such that the condition is satisfied. A consequence interpolation is that the cyclic extension appended to the image must be increased by half of the size of the interpolation filter memory.

7.5.3   Capacity Estimate

As is the case in any practical system, the MIMO wireless optical channel is both peak- and average-amplitude limited. A closed form expression for the capacity of such links is not yet know, however, we resort to approximating the capacity using an additive white Gaussian noise channel, as was done in Section 7.4. Define an electrical power constraint, $E = KP^2$ for average optical power constraint, P and constant $K > 0$. The DC spatial frequency bin of every transmitted frame is set to be the constant,

$$A(0,0) = n_{\text{T}}P,$$

so that the average amplitude constraint is satisfied for every frame. This assumption ensures that the average optical power constraint is met, however, it is pessimistic since it has been demonstrated in single element systems that varying the optical bias per symbol can lead to significant gains in OSNR [104]. The constant K is chosen to ensure that the clipping distortion of the channel is minimized. Any remaining clipping distortion is considered as noise in the power allocation. In Section 8.4, a specific value of K is chosen for an experimental prototype setup.

Shot-noise, electronics noise as well as spatially aliased noise due to the clipping and quantization distortion contribute to the noise in each spatial frequency bin. These noise components are signal-dependent and correlated. In Section 8.2, the signal dependent noise is characterized over a number of spatial frequency bins. In order to estimate the capacity, assume that the noise variance in each bin, $\sigma^2(u, v)$, is the maximum possible over all input signals. This is a pessimistic assumption, however, it models the noise in each bin as signal independent and allows for analysis.

If a cyclic extension is appended and spatial synchronization is performed, as discussed in Section 7.5.2, the channel is a series independent Gaussian spatial frequency channels. The capacity of this electrical channel is well known to be [128],

$$C = \sum_{u,v} \frac{1}{2} \log \left(1 + \frac{E(u,v)}{\sigma^2(u,v)} \right) \tag{7.9}$$

where the power allocation per spatial frequency bin is

$$E(u,v) = \begin{cases} \nu - \frac{\sigma^2(u,v)}{|H(U,v)|^2} & \text{if } \nu > \frac{\sigma^2(u,v)}{|H(U,v)|^2}, \\ 0 & \text{otherwise} \end{cases} \tag{7.10}$$

for a constant ν such that

$$\sum_{u,v} E(u,v) = KP^2. \tag{7.11}$$

The optimal power allocation over the spatial frequency bins is arrived at by "water-pouring" the electrical power constraint over the bins to occupy those spatial frequency bins with the lowest $\sigma^2(u,v)/|H(u,v)|^2$. Again, this result is analogous to results in MIMO RF channels, in which the power constraint is distributed over the singular values of the channel matrix [189, 190]

Since $H(u,v)$ will vary depending on the link configuration, it is difficult to draw general conclusions about the capacity of the MIMO wireless optical channel. Consider the instructive but unrealizable example of a flat, strictly spatially bandlimited channel

$$H(p,q) = \begin{cases} 1 & \text{if } p \in [-W_S, W_S], q \in [-W_S, W_S]; \\ 0 & \text{otherwise,} \end{cases}$$

where $\sigma^2(u,v)$ is constant over all frequency bins. The power allocation (7.10) for this optical transfer function will distribute the power equally over all bins. Notice then, that this example is a pixel-matched channel in spatial frequency domain. The number of spatial channels available in the channel is limited by n_T, n_R and the spatial bandwidth of the channel. Consider the case when the spatial bandwidth of the channel is less than the Nyquist frequency of the receive array such that,

$$W_S < 2\pi/D_R.$$

A channel in this regime of operation is termed *spatially bandwidth limited*, since for a fixed imaging array size, increasing the number of receive elements such that

$$n_{Rx} > \frac{W_S n_{Rx} D_R}{2\pi} + 1 \quad \text{or} \quad n_{Ry} > \frac{W_S n_{Ry} D_R}{2\pi} + 1$$

does not add any new spatial channels within the Nyquist region. Conversely, if

$$W_S \geq 2\pi/D_R$$

then the channel is termed *pixel limited*, since increasing n_R increases the number of spatial channels.

As the link distance reduces, the OSNR and the spatial bandwidth of point-to-point pixelated channels increase as can be seen in (7.3). The design goal is, for a given distance, to increase the number degrees of freedom until the channel enters the spatially bandwidth limited regime. Short-range pixelated optical channels provide significant improvements in spectral efficiency, as is the case with pixel-matched systems, by fully exploiting the spatial degrees of freedom provided by the channel. However, unlike pixel-matched systems, the pixelated channel takes into account the spatial frequency response of the channel in the design of signalling. It is important to note that these gains in spectral efficiency are only significant in channels with sufficient OSNR such as in interconnection and short-range links.

7.6 Conclusions

The MIMO wireless optical channel exploits spatial degrees of freedom inherent in the channel to provide gains in spectral efficiency. Pixel-matched systems provide a spatial multiplexing gain at the expense of requiring that all transmitters and receivers be in perfect alignment. The pixelated wireless optical channel with SDMT spatial modulation provides a means to generate spatial modulation which is well suited to the spatial frequency constraint. Additionally, the use of these bases eliminates the need for exact spatial alignment of the transmit and receive pixels and requires only that the spatial frequency resolutions be matched.

The gain in using MIMO wireless optical links over single element links arises only at high optical SNRs. Potential applications for the pixelated wireless optical channel are in the areas of high-speed, short range links where this high SNR condition exists. One example is optical backplane applications connecting a number of circuit boards. High OSNR wireless channels also exist in a host of inter-device connection applications to provide a high-rate, secure, fixed wireless link between a portable computer and host network. For example, a pixelated wireless optical link could be designed to improve the data rate of present day links to all for the transfer gigabytes of data, video and high-quality audio to hand-held computing devices, while providing a secure link free of mechanical wear and interface issues.

In next chapter, an experimental channel is characterized to produce a channel model for computer simulation studies. The capacity of a number of spatial bases is computed for uncoded and SDMT modulated spatial signalling.

Some candidate spatio-temporal coding schemes are considered for the pixelated wireless optical channels to quantify the gains available by using this channel topology.

Chapter 8

PROTOTYPE MIMO OPTICAL CHANNEL: MODELLING AND SPATIO-TEMPORAL CODING

The previous chapter introduced the pixelated wireless optical channel as a space-time link which exploits spatial diversity to achieve increases in spectral efficiency. Multiple transmitter and receiver space-time codes have been proposed for radio frequency channels to improve spectral efficiency [196–198]. These codes, however, are not directly applicable to the optical intensity channel. As a result, the channel model for the pixelated optical channel differs fundamentally from the radio channel.

This chapter describes the construction of an experimental prototype point-to-point pixelated wireless optical channel [4]. Channel measurements are presented along with a channel model amenable to computer simulation. This chapter also presents signalling and coding strategies for the pixelated wireless optical channel which demonstrate the potential of this channel topology to achieve high spectral efficiencies. This chapter considers SDMT modulation and estimates the capacity of a given channel realization by way of the water-pouring spectrum. Multi-level coding coupled with multi-stage decoding is also implemented in simulation to approach the capacity with realizable algorithms. The resulting communications link achieves rates of approximately 17.1 kbits per frame in simulation which is 76% of the estimated channel capacity.

The pixelated wireless optical channel is thus demonstrated to be a space-time channel capable of achieving high spectral efficiencies using available coding techniques.

8.1   Experimental Prototype

In order to determine the nature of the pixelated optical channel, an experimental point-to-point link, as shown in Figure 7.1, was constructed. The constructed link operates in the visible range of wavelengths using commercially available components. The goal of this study is to gain insight into the impairments of the channel as well as issues involved in the design of such links. This section describes the components of the link as well as the link configuration.

8.1.1   Transmit Array

The transmit array was realized using the LCD panel of an NEC Versa 6050MX laptop computer. The LCD panel measures 12.1" on the diagonal with 1024×768 pixels at a dot pitch of 0.24 mm. The video card is based on the Chips & Technologies 65550 chip set with 2MB of video RAM. The CPU is an Intel Pentium 150 MHz processor with 32 MB of system memory [199]. The laptop computer was used to generate test data and control the LCD panel for the purposes of implementing the link.

In order to implement control of the display, a graphics driver was written for the laptop video hardware. The channel response under a variety of transmitted optical intensities is characterized by transmitting static frames as well as a series of frames at a given frame rate, as described in Section 8.2.

8.1.2   Receive Array

The array of light receivers is implemented by way of the Basler A301f CCD camera [200]. Typical applications for this camera are for manufacturing inspection and quality control. This camera has a resolution of 640x480 pixels of size $10 \mu m \times 10 \mu m$ and can operate at frame rates of 60 frames per second. The camera outputs an uncompressed 8-bit grey-scale value for each pixel in the array. The camera interfaces to the external devices via an IEEE 1394 (Firewire) interface. This inexpensive industry standard interface provides a peak rate of 400 Mbps out of the camera.

Coupled to the camera is a fixed-focal-length lens manufactured by Tamron Inc. [201]. This lens has a fixed focal length of 25 mm. Typical applications for this device are for closed-circuit television and some machine vision systems.

An IEEE 1394 PC card was installed in a 1 GHz Pentium III desktop computer with 256 MB of memory to interface with the camera. Software was then written on the host to control the camera and to store the received frames. Interface to the camera was accomplished using the Matrox Image Library (MIL-Lite) version 7.1 [202]. These routines allowed for camera initialization and interface. Grabbed frames were transmitted to the host computer and stored in host memory. Once a number of frames was acquired, the frames were saved to the hard drive as bit-mapped images. Processing of the images was performed off-line using Matlab [203].

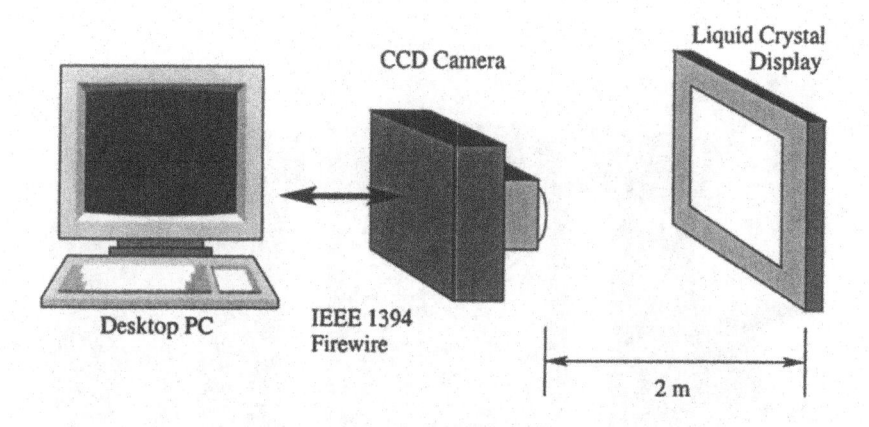

Figure 8.1. Diagram of experimental prototype.

8.1.3   Prototype Configuration

Figure 8.1 pictorially depicts the configuration of the experimental prototype link. The channel is configured so that the optical axis of the transmitter and the receiver are aligned. In this case, the received image is an orthographic projection of the transmitted image, as discussed in Section 7.2.4. The distance between transmitter and receiver was set at 2 m and no relative motion between ends of the link was present. The lens was manually focused on the LCD panel and the aperture was set to f/1.6 for all measurements to minimize detector saturation.

The transmit image is assumed to be divided into two regions: the data block and the timing bars. Figure 8.2 presents a photograph of a typical transmit frame in the case of a binary level modulator. The data block is a 512×512 set of transmit pixels which form the data-bearing portion of the transmitted image. The timing bars on either side of the data block change from the maximum to the minimum intensity at each frame transmission. In this manner the frame clock is transmitted to the receiver. As discussed in Section 7.3.4, conventional timing recovery techniques can be applied to these timing bars once the registration has been acquired.

At the receiver an intensity image is formed which is a scaled version of the transmitted image corrupted by spatial ISI and noise. There is no temporal ISI in this experimental channel since there is not multipath component and the frame rate was far below the bandwidth of the electronics. At the receiver, the sampled intensity image has dimensions 154×154 pixels. Figure 8.3 presents a comparison of a transmitted binary spatial repetition coded test frame, described in Section 8.3, along with the corresponding received image. Notice that the received image is distorted relative to the transmit image due to the channel low pass spatial frequency response as well as noise sources. It is the task of the

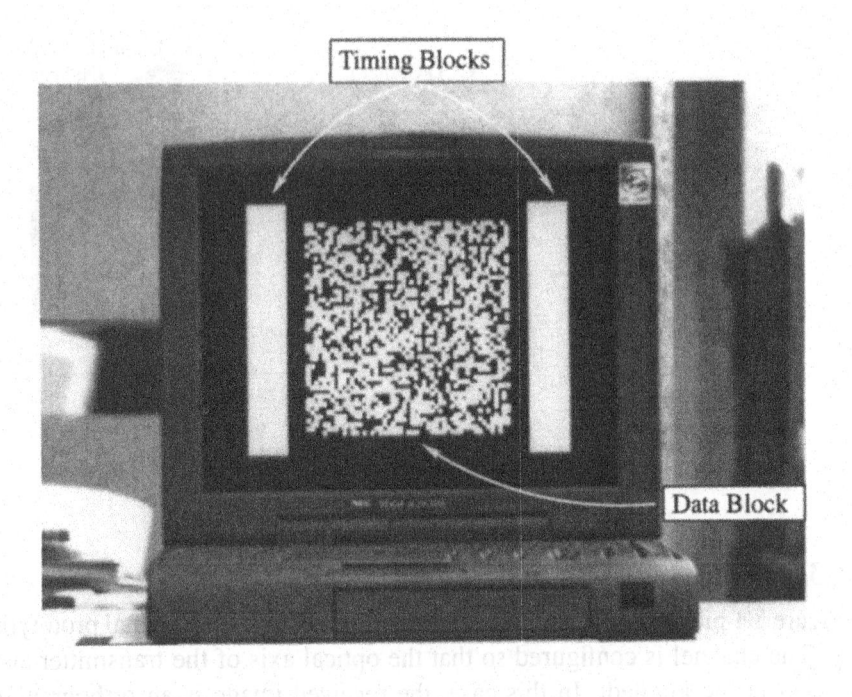

Figure 8.2.   A captured image of a single frame of binary level test system.

Figure 8.3.   Example of a transmitted frame and corresponding measured received frame.

receiver to synchronize itself in time and space to the transmitted image and to detect the received data block image. The channel model presented in the next section is developed to characterize these impairments to allow for signalling design.

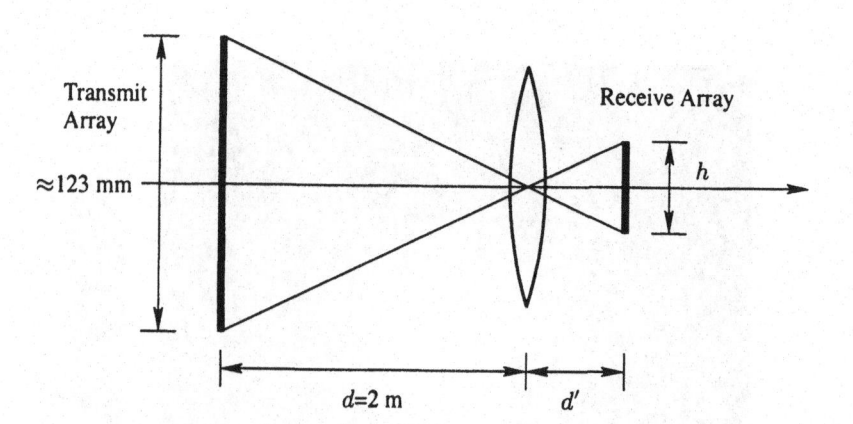

Figure 8.4. Estimating of size of receive array from the geometry of the configuration.

8.2 Channel Model

This section describes a series of measurements made on the experimental channel of Section 8.1.3. The ultimate goal of these measurements is to produce a channel model which is amenable to computer simulation and which provides insight into the channel impairments present.

8.2.1 Point-Spread Function

The array is composed of 512×512 pixels centered in the middle of the LCD panel. At the receiver, the size of the receive image was measured to be 154×154 pixels. Figure 8.4 illustrates the geometry of the link. The transmitter array has 512 elements per side and the dot pitch is 0.24 mm, as described in Section 8.1.1, which gives an approximate size of 122.86 mm per side.

Using the thin-lens equation [185] with focal length $f = 25$ mm,

$$\frac{1}{f} = \frac{1}{d} + \frac{1}{d'}$$

to give $d' = 25.3$ mm. Using similar triangles and the geometry of Figure 8.4 gives

$$\frac{122.86}{2000} = \frac{h}{25.3}$$

which gives $h = 1.55$ mm. Since the size of each receive pixel is approximately $10\mu m \times 10\mu m$, as described in Section 8.1.2, the number of receive pixels per side is approximately 155 which is close to the measured value of 154.

In order to characterize the response of the link, point sources at maximum intensity were transmitted at five locations in the array. Five point-spread functions were measured due to single illuminated pixels at transmitter pixel locations (256,256), (128,128), (128,384), (384,128), (384,384) and are termed

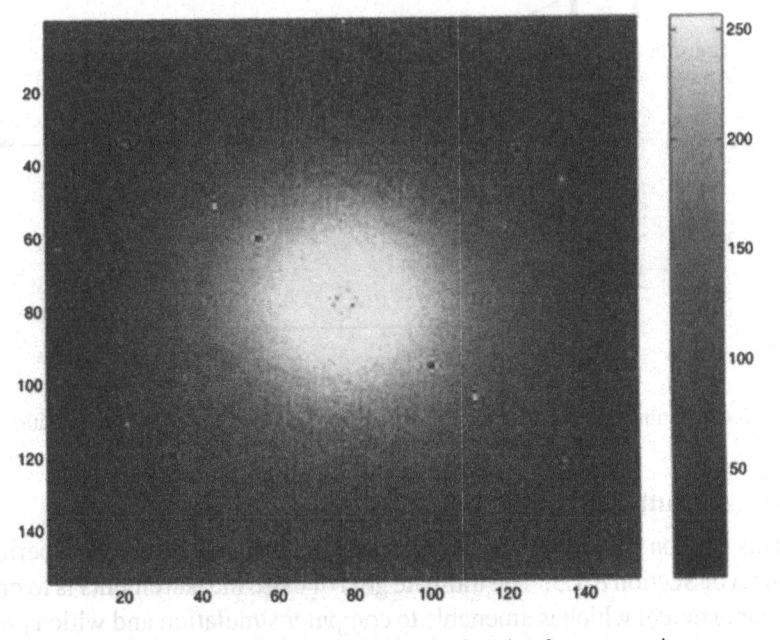

Figure 8.5. Magnitude of Optical Transfer function for center point.

center, top left, top right, bottom left and bottom right respectively. For each response, 10000 frames were averaged to produce the resulting point-spread function. The zero-input response of the channel was also sampled and averaged over 10000 frames and subtracted from the point-spread functions. This was done to reduce the offset caused by glare or systematic spatial variation across the LCD panel. Figures 8.5 and 8.6 present plots of the average magnitude of the optical transfer functions for these five cases.

A common assumption present in most optical systems is that the point-spread function is circularly symmetric [185]. In this special case the Fourier transform is known as the *Hankel transform* and the optical transfer function is real and circularly symmetric. Under this assumption, the measured point-spread functions should be real with some complex exponential term accounting for the spatial shift. Assuming that the spatial frequency response is real and non-negative, taking the magnitude of the received responses removes this spatial shift component allowing for comparison of the responses.

Table 8.1 presents the percent difference in energy between the central pulse and the corner pulses. By this metric there is some spatial dependence in the impulse responses, however, the assumption of spatial invariance hold approximately in this case. Since the variations are relatively small the channel is modelled as being spatially invariant.

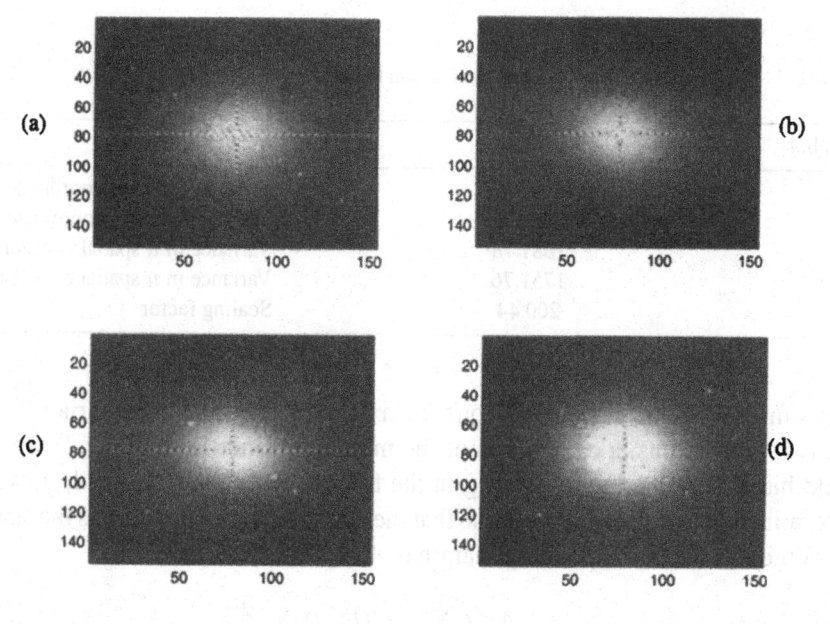

Figure 8.6. Magnitude of Optical Transfer function for (a) top left, (b) top right, (c) bottom left, (d) bottom right.

Table 8.1. Comparison of percent difference in energy between magnitudes of corner optical transfer functions and center.

Top Left	8.6 %
Top Right	7.8 %
Bottom Left	4.0 %
Bottom Right	2.1 %

The sampled central optical transfer function, $P_C(k, l)$, is then taken as the prototypical response of the channel. A common optical transfer function, $G(u, v)$, used in the characterization of some optical systems is the two-dimensional Gaussian pulse [186]

$$G(u, v) = K \exp\left(-\frac{(u - \mu_u)^2}{2\sigma_u^2} - \frac{(v - \mu_v)^2}{2\sigma_v^2}\right) \tag{8.1}$$

for some constant K. The Gaussian response was fit to $P_C(k, l)$ by computing the center of mass of $P_C(k, l)$ in the u coordinate as

$$\mu_u = \frac{\sum\sum k P_C(k, l)}{\sum\sum P_C(k, l)}$$

Table 8.2.   Parameters of optical transfer function fit.

Symbol	Value	Meaning
μ_u	76.74	Mean in u spatial coordinate
μ_v	76.67	Mean in v spatial coordinate
σ_u^2	1681.78	Variance in u spatial coordinate
σ_v^2	1751.76	Variance in u spatial coordinate
K	200.44	Scaling factor

and similarly in the v spatial frequency axis. In order to fit the variances in the two dimensions, a calculation of the moment of inertia of $P_C(k, l)$ would yield biased results due to aliasing in the frequency domain. Instead, σ_u was determined by assigning its value so that the u-marginal of $G(u, v)$ has the same amplitude as the corresponding marginal of $P_C(k, l)$, that is,

$$\sigma_u^2 = \frac{1}{2\pi} \left(\frac{\sum_l P_C(k^*, l)}{\sum \sum P_C(k, l)} \right)^{-2}$$

where k^* is the closest value of k to μ_u. An analogous computation was carried out for the other spatial frequency dimension. The value of the constant K can then be determined by ensuring that the volumes under $G(u, v)$ and $P_C(k, l)$ are the same to give

$$K = \frac{\sum \sum P_C(k, l) \Delta^2}{2\pi \sigma_u \sigma_v}$$

where Δ^2 is volume associated with each sample point. Table 8.2 presents the parameters of the fit. Note that μ_u and μ_v are taken with respect to the discrete spatial frequency domain and correspond to the DC value. Thus the optical transfer function can be approximated by a Gaussian pulse about the origin. Figure 8.7 shows the resulting fit for a given cross-section of $P_C(k, l)$ and $G(u, v)$.

With a closed form expression for the optical transfer function, the inverse Fourier transform of $G(u, v)$ was taken analytically to obtain the point-spread function,

$$g(x, y) = K \frac{\sigma_u' \sigma_v'}{2\pi} \exp \left(-\frac{x^2}{2/\sigma_u'^2} - \frac{y^2}{2/\sigma_v'^2} \right) \qquad (8.2)$$

where

$$\sigma_u' = \sigma_u \frac{2\pi}{154} \quad \text{and} \quad \sigma_v' = \sigma_v \frac{2\pi}{154}$$

are the normalized variances in spatial frequency domain.

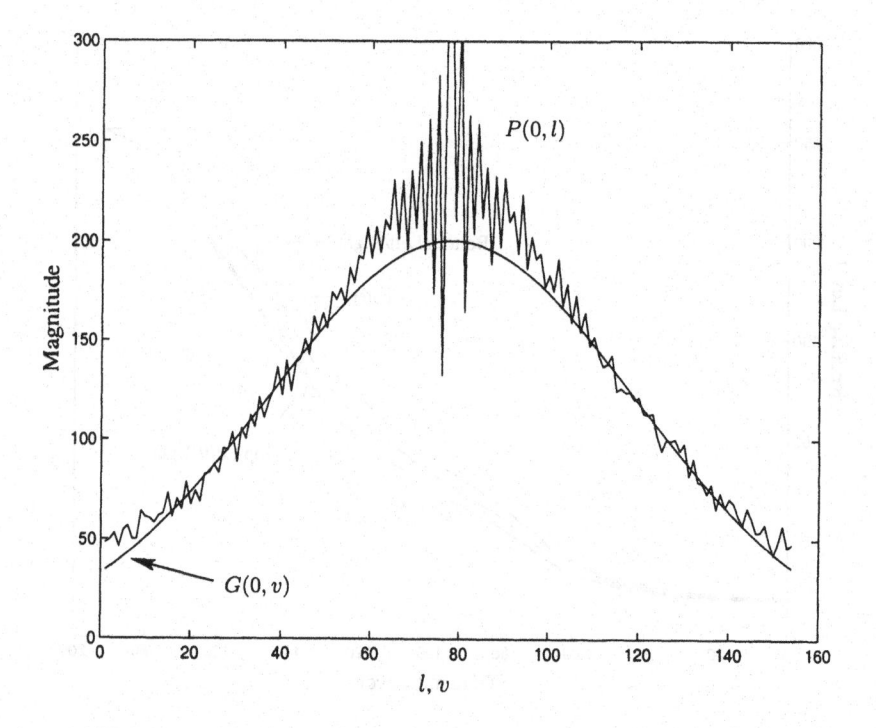

Figure 8.7. Gaussian fit to $P_C(k,l)$ for a single cross-section for $k = 0$, $u = 0$.

8.2.2 Non-Linear Distortion

In order to measure the gamma distortion for the display used in the link, the transmit image was set to be a constant grey value between 0 and 255. The noise at the receiver can be assumed to be zero mean without loss of generality since any non-zero mean will be incorporated into the gamma distortion measurement. The receive images are averaged over 1000 frames to produce an estimate of the received intensity value in the absence of noise. Figure 8.8 presents the output 8-bit value received as a function of the 8-bit grey-scale value transmitted. The gamma distortion is measured for five pixel locations in the 154×154 element receive array: (38,38), (38,117), (77,77), (117,38), (117,117), termed top left, top right, center, bottom left and bottom right respectively. Notice that the non-linear distortion is a function of position. It is also important to note that for input values greater than 190, the output received level is saturated to 255 by the quantizer at the receiver.

8.2.3 Noise

The temporal noise source in each pixel is due to a wide variety of physical phenomena. The distribution of this noise is typically taken as being Gaussian and signal dependent [182]. Figure 8.9 presents normalized histograms for the

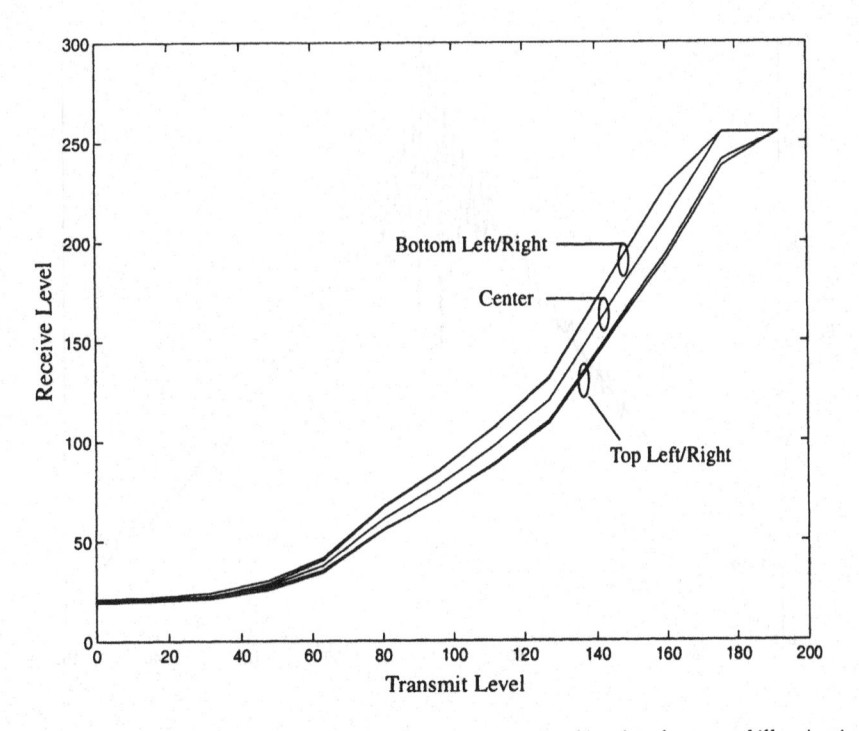

Figure 8.8.   Receive average amplitude level versus transmitted level under normal illumination.

receive levels for a variety of transmit levels. The measurements were done by transmitting a constant value at each pixel of the LCD. The histogram presented is for the center pixel of the receive array and normalized so that the sum of the frequency observations is unity. Both the mean and variance of the observed signals are clearly signal dependent and both increase as does the transmit level.

These histograms were collected for the five locations on the receive array under conditions of low and normal background illumination. The low illumination state was achieved by performing the measurements in a room where all illumination was turned off and external light sources minimized. Normal illumination refers to measurements taken when background illumination was present. The measurements were done in these two environments to determine the impact of the signal dependent component of the noise.

Under conditions of low illumination the histograms were collected for the five sample points over 1000 frames. The sample average was computed from these measurements and is presented in Section 8.2.2. This average received value is proportional to the received optical intensity as discussed in Section 2.1.2. The sample variance was also computed and plotted versus the received optical intensity (i.e., the mean value) in Figure 8.10. Note that at high received values, the quantizer saturates the receive value and the variance goes to zero. The noise distribution is conventionally modelled as being Gaussian

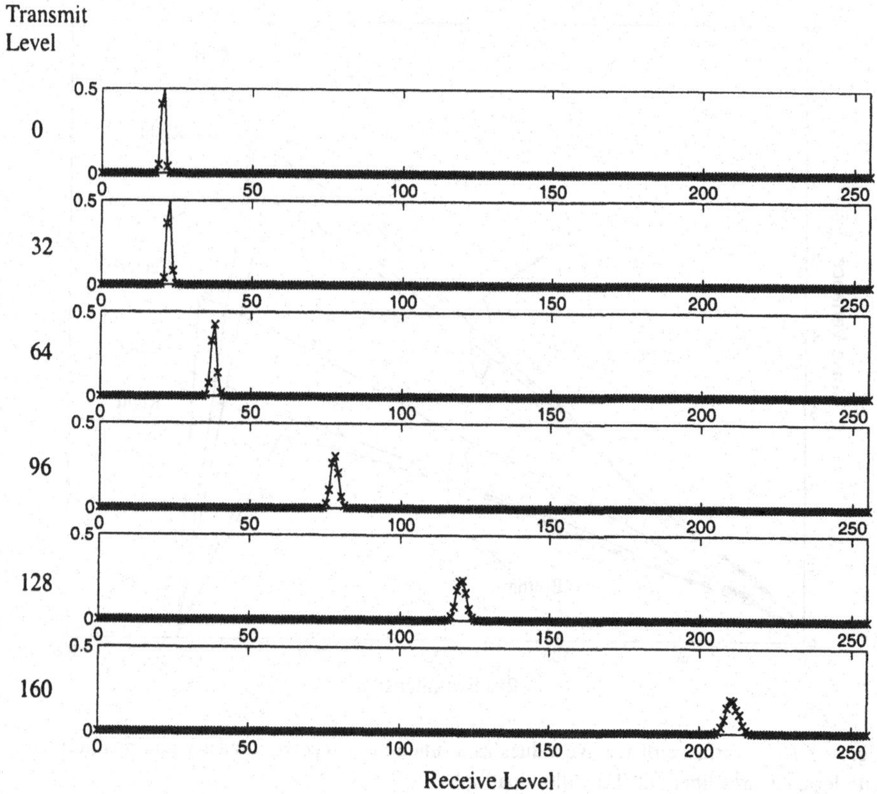

Transmit
Level

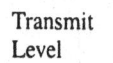

Receive Level

Figure 8.9. Estimate of the distribution of receive values over a range of transmit levels.

with variance linearly related to the receive optical intensity [182]. The signal dependence can be justified by noting that each receive pixel is in essence a narrow field-of-view device where the impact of the receive optical intensity on the noise can not be neglected, as described in Section 2.3. The variance of the noise is fit to the linear model

$$\sigma^2 = \alpha \bar{I} + \beta \qquad (8.3)$$

where $\bar{I}$ is the mean received level. The parameter α quantifies the signal dependence of the noise while β is the background noise term used thus far. As a result, the previous case of signal independent noise is a special case of this model. Table 8.3 presents the linear fit values as well as the R^2 statistic of fit, where R^2 close to 1 indicates a good fit to the linear model [204]. An identical procedure was carried out in the case of normal background illumination and the results are presented in Figure 8.11 and Table 8.4.

In both cases, there is significant spatial variation in the received variance values. A linear model for the variance variation is also appropriate due to the relatively good fit in the two cases. It should be noticed, however, that

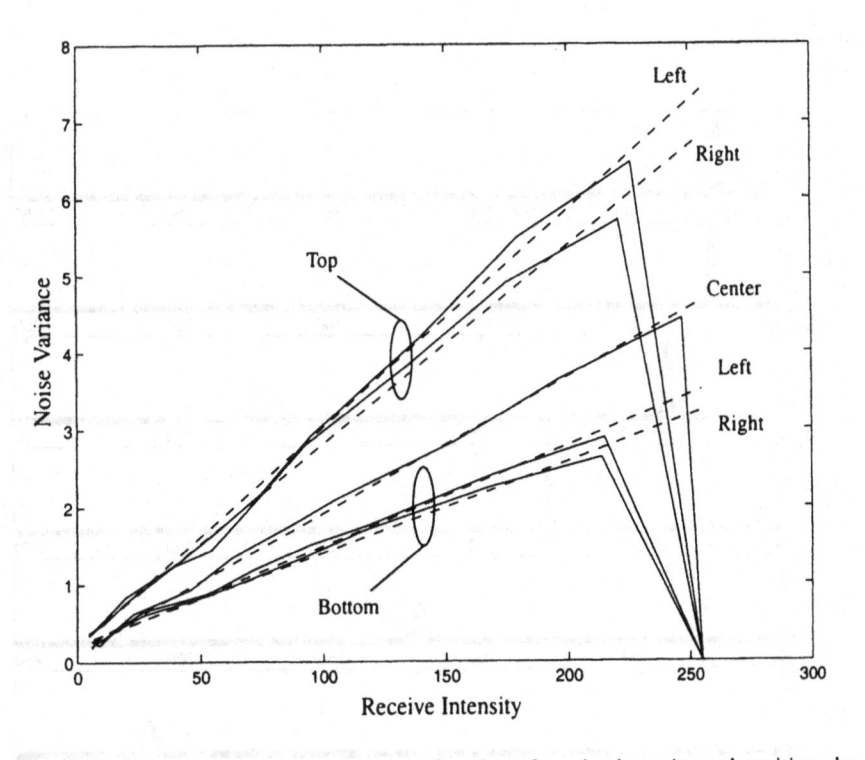

Figure 8.10. Variance of receive values as a function of receive intensity and position along with least squares linear fit (Low illumination).

Table 8.3. Variance fit parameters along with R^2 statistic of fit (Low illumination).

	α	β	R^2
Top Left	0.0282	0.2080	0.9977
Top Right	0.0257	0.2418	0.9960
Center	0.0176	0.1581	0.9981
Bottom Left	0.0130	0.1938	0.9935
Bottom Right	0.0118	0.2278	0.9949

in the case of low and normal illumination the values for signal dependence parameter α in (8.3) do not change a great deal, whereas the values of β change significantly. Thus, the impact of ambient light on noise variance is captured in the β parameter while the signal dependence of the noise is captured in the α parameter. In single-element links with signal independent noise, considered in earlier chapters, the α term is modelled as being insignificant with respect to the β parameter. In the narrow field-of-view case present in the pixelated wireless optical channel, this assumption is no longer valid.

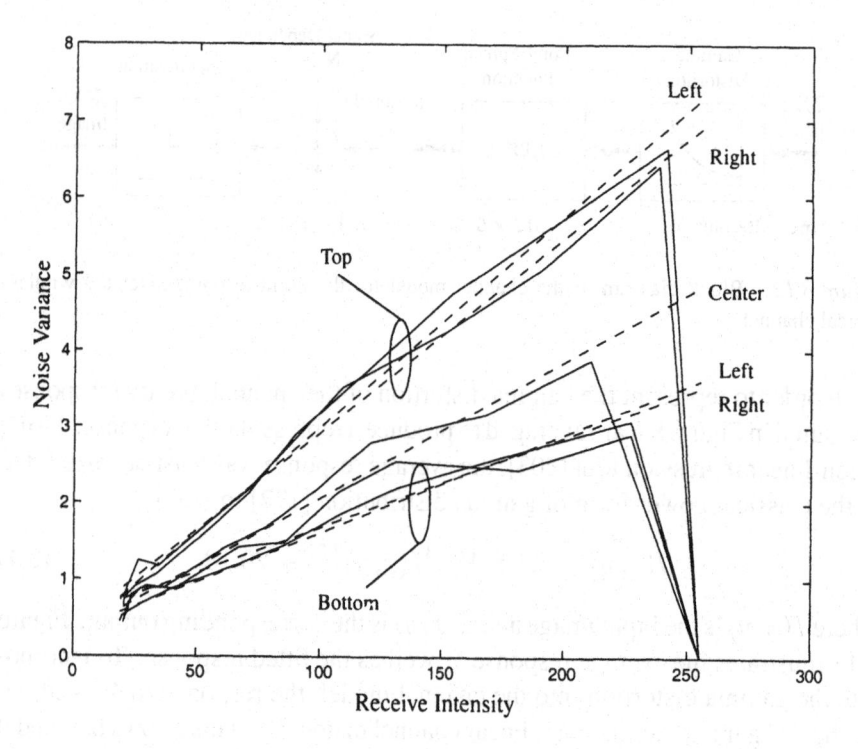

Figure 8.11. Variance of receive values as a function of receive intensity and position along with least squares linear fit (Normal illumination).

Table 8.4. Variance fit parameters along with R^2 statistic of fit (Normal illumination).

	α	β	R^2
Top Left	0.0277	0.2250	0.9948
Top Right	0.0256	0.3616	0.9921
Center	0.0180	0.2657	0.9801
Bottom Left	0.0129	0.3208	0.9879
Bottom Right	0.0118	0.4981	0.9134

8.2.4   Simulation Model

Figure 8.12 presents a block diagram of the simulation model developed for this point-to-point pixelated wireless optical channel.

The input image is at a resolution of 512×512 pixels which corresponds to the transmit array. The input image is assumed to be a grey-scale image with input values between 0 and 255.

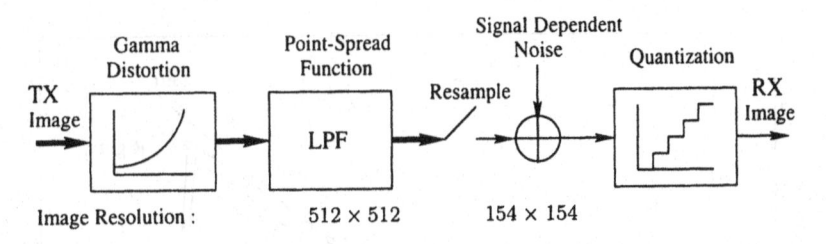

Figure 8.12. Block diagram of the channel model for the point-to-point pixelated wireless optical channel.

In order to represent the gamma distortion of the channel, the five responses measured in Figure 8.8 are averaged to produce a representative response. Using a non-linear regression tool [203], the average response was least-squares fitted to the classical power form of gamma distribution [177] to yield

$$J(x,y) = 1.1 \times 10^{-3} I(x,y)^{2.37} + 20.18 \qquad (8.4)$$

where $I(x,y)$ is the input image and $J(x,y)$ is the corresponding output. Figure 8.13 illustrates the average response as well as the fitted response. To incorporate the gamma distortion into the channel model, the response (8.4) is scaled by the DC gain of the low-pass linear channel distortion. This allows the model to reflect the measured response accurately.

The point-spread function (8.2) is sampled at the high spatial rate of the transmitted 512×512 pixel image to yield a point-spread function to be used for simulation. The resulting upsampled point-spread function is shown in Figure 8.14 and is used in the simulation model. The filtering is done in spatial domain using the upsampled point-spread function.

The resulting image is resampled to a resolution of 154×154 using bilinear interpolation to determine non-integer sample points [194].

The noise is modelled as being independent both in space and in time. This is taken as a worst case model for the array since any correlation in the noise can be exploited to improve transmission. Additionally, the noise is modelled as being zero mean. The average value of the α and β parameters in (8.3) are taken over all cases of normal illumination to be $\bar{\alpha} = 0.0192$ and $\bar{\beta} = 0.3343$ for every pixel in the receive array. The received signal is then quantized to the integer values between 0 and 255.

A comparison between 100 transmitted frames constructed with the binary modulation described in Section 7.3 versus the output of the model gives rise to a 5.4% error in terms of the energy of the difference divided by the energy of the channel measurement. Although not insignificant, the error between the measured data is small enough to suggest that the salient properties of the channel are represented in the model and that the model is appropriate for modem design purposes.

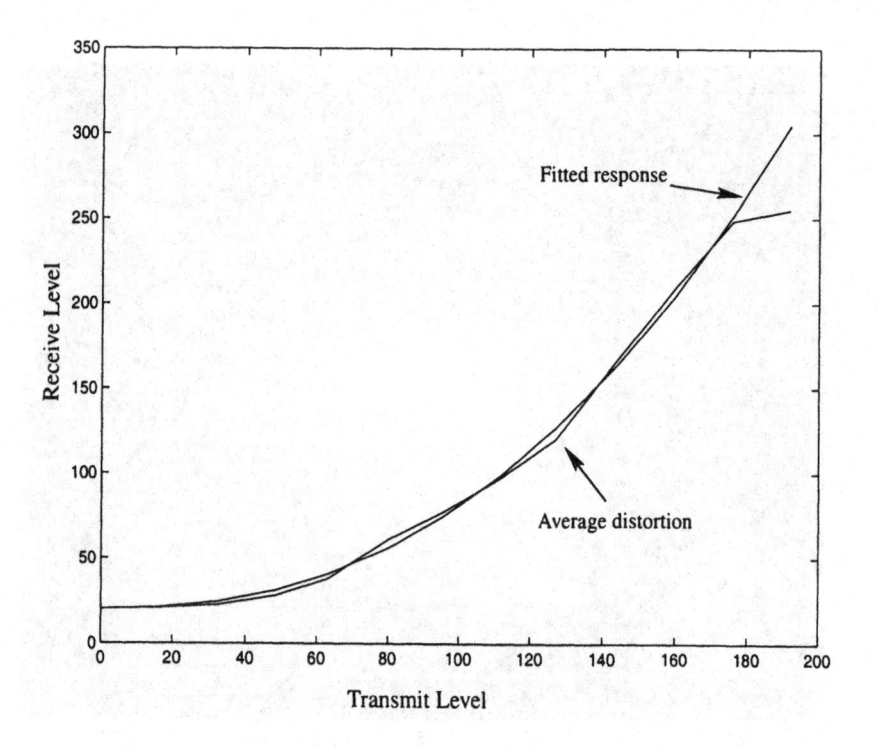

Figure 8.13. Average gamma distortion along with least squares fit.

Table 8.5. Parameters of the channel model for the prototype link.

	Parameter	Value
	μ_u	76.74
Point-Spread	μ_v	76.67
Function (8.1)	σ_u^2	1681.78
	σ_v^2	1751.76
	K	200.44
Signal Dependent	α	0.0192
Noise (8.3)	β	0.3343
Non-Linear Gamma	γ_1	1.1×10^{-3}
Distortion (8.4)	γ_2	2.37
	γ_3	20.18

The parameters used in the defining the simulation model are summarized in Table 8.5.

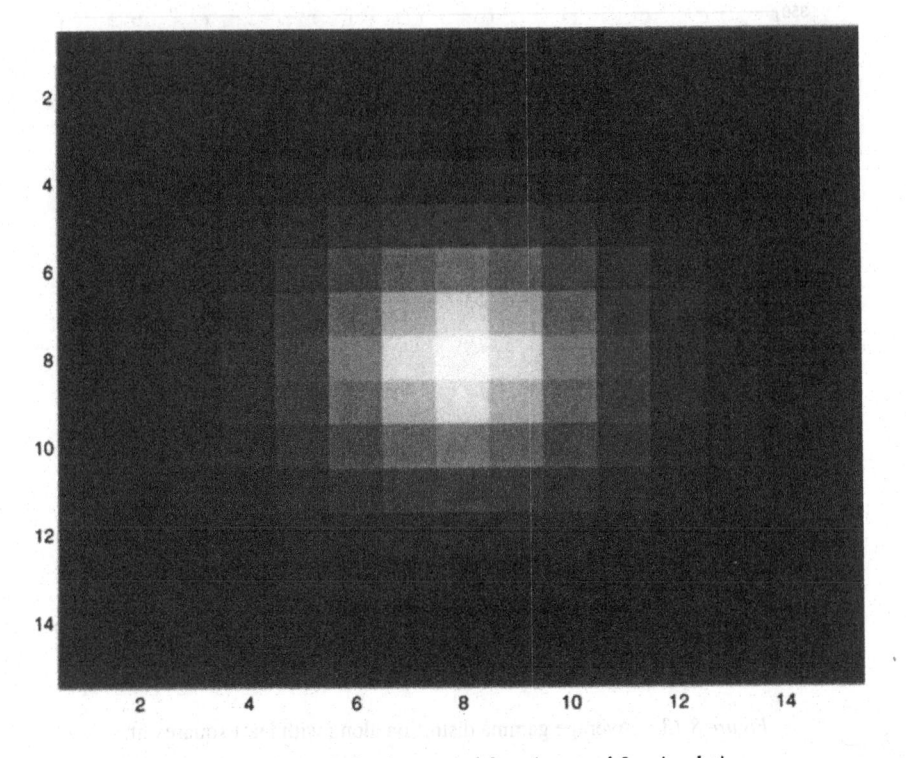

Figure 8.14. Resampled point-spread function used for simulation.

8.3   Pixel-Matched Systems

As discussed in Section 7.4, pixel-matched systems implement a simple spatial signalling technique in order to realize a multiplexing gain. However, in the previous analysis the impact of the spatial response of the channel was not considered in the performance of such systems.

Consider implementing a pixel-matched spatial modulation scheme by partitioning the transmit array into a series of square regions, termed macropixels, each with an integral number of transmit pixels of the LCD display. Each transmit pixel assigned to a given macropixel transmits an identical PAM signal at the frame rate. Thus, a spatial repetition code is employed in every macropixel. Figure 8.3 presents an example of such a system in which macropixels consist of 10×10 arrays of transmit pixels and binary level PAM is used for each macropixel.

The analysis in Section 7.4 estimated the increase in spectral efficiency in pixel-matched systems in which the spatial response of the channel was flat over all frequencies. As discussed in Section 7.3.1, the pixelated wireless optical channel has a low pass spatial frequency response. As a result, the assumption that adjacent pixels are independent channels becomes increasingly worse as the macropixel size decreases. In order to provide a more realistic view of the impact of spatial frequency response on the transmission rate, the capacity was estimated when binary level images were transmitted through the channel model derived in Section 8.2 for a variety of macropixel sizes. In each case independent, equally likely data was transmitted in each macropixel and perfect spatial registration was assumed. Since spatial synchronization is not possible, spatial matched filter detection would require that a filter be designed for each macropixel. Matched filter detection was approximated by using a single response and linearly interpolating between the samples of the filter output to provide the correct sampling instant. An approximated Wiener spatial equalizer [194] was implemented on the matched filter outputs by assuming the noise and signal power spectral densities are white and by measuring the variance over a number of test frames. Figure 8.15 presents normalized histograms representing the distribution of the receive amplitudes given a zero or a one is transmitted in the case of 10×10 sized macropixels. The figure also presents a Gaussian fit to the histograms. Using the Gaussian fit to the conditional distributions, the capacity of each macropixel channel is estimated using a well known numerical technique [142]. Figure 8.16 plots the capacity of the pixelated link in units of bits per frame for a number of macropixels. Notice that, as in Figure 7.4, the rate of the channel increases with the number of macropixels. However, in this case, as the number of macropixels increases further, i.e. as the size of each macropixel decreases, the residual ISI remaining in the received signal reduces the transmission rate of the link. In this simulation example, the maximum data rate is less than 3 kbits/frame.

Pixel-matched systems for the point-to-point wireless optical channel are simple to construct but suffer from several key disadvantages. Sending independent data in each macropixel does not take into account the spatial frequency response of the channel and suffers from aliasing distortion. The lack of spatial synchronization makes detection and equalization difficult since filters must ideally be designed for each macropixel. In spite of these difficulties, simulations indicate that this spatial modulation still provides gains in spectral efficiency over single element links which are limited by the spatial response of the channel. The following sections use SDMT based spatio-temporal coding which takes into account the spatial frequency response of the channel to further improve the spectral efficiency further.

8.4 Pixelated Wireless Optical Channel

In this section, the capacity of the point-to-point wireless optical channel with SDMT spatial modulation, presented in Section 7.5, is estimated using the derived channel model. In order to investigate the feasibility of achieving this promised rate, multi-level codes as well as multi-stage decoding are applied to

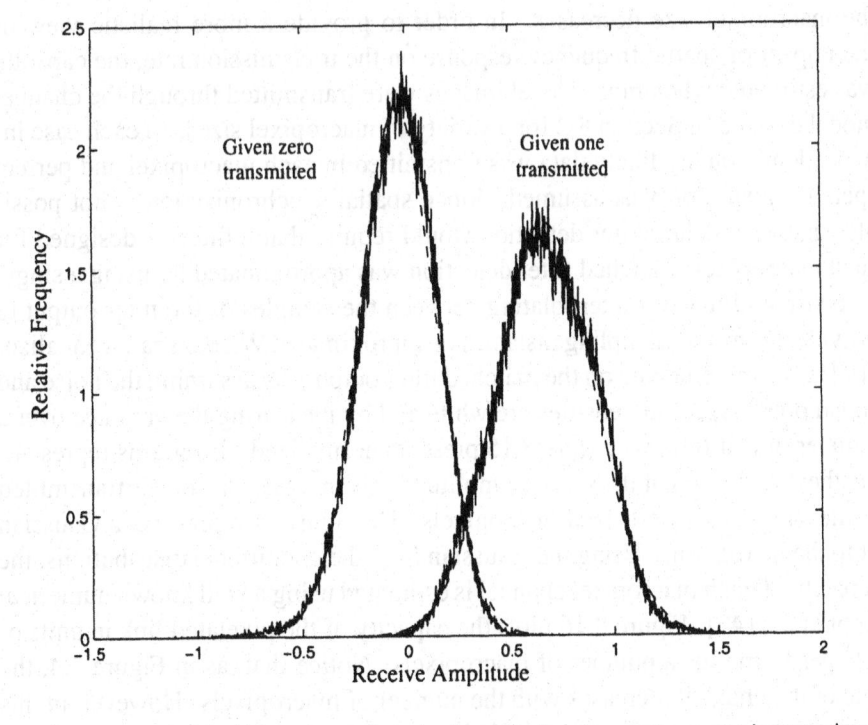

Figure 8.15. Normalized histograms of received amplitudes given a zero or one is transmitted for 10 × 10 sized macropixels along with Gaussian density fit.

the channel to provide a realizable system which, in simulation, achieves rates of 76% of the channel capacity.

8.4.1   System Overview

The system model for the SDMT channel is identical to that of Figure 7.5 with $n_{Tx} = n_{Ty} = 512$ and $n_{Rx} = n_{Ry} = 154$. It is assumed that the transmitter is able to pre-distort the transmitted image in order compensate for the gamma distortion of the display. Additionally, it is assumed that spatial registration has taken place, i.e., the receiver knows the coordinates of the transmit pixels in the receive plane.

In order to compute a power allocation, the receiver must feed back the channel characteristics to the transmitter. A simple technique to accomplish this is to fit the channel point-spread function to a family of Gaussian functions represented by the parameters K, μ_u, μ_v, σ_u^2 and σ_v^2 defined in Section 8.2.1. The receiver can then feed back the point-spread function parameters along with an estimate of the worst case noise to the transmitter using simple and robust on-off keying. With this information, the transmitter can compute a power allocation

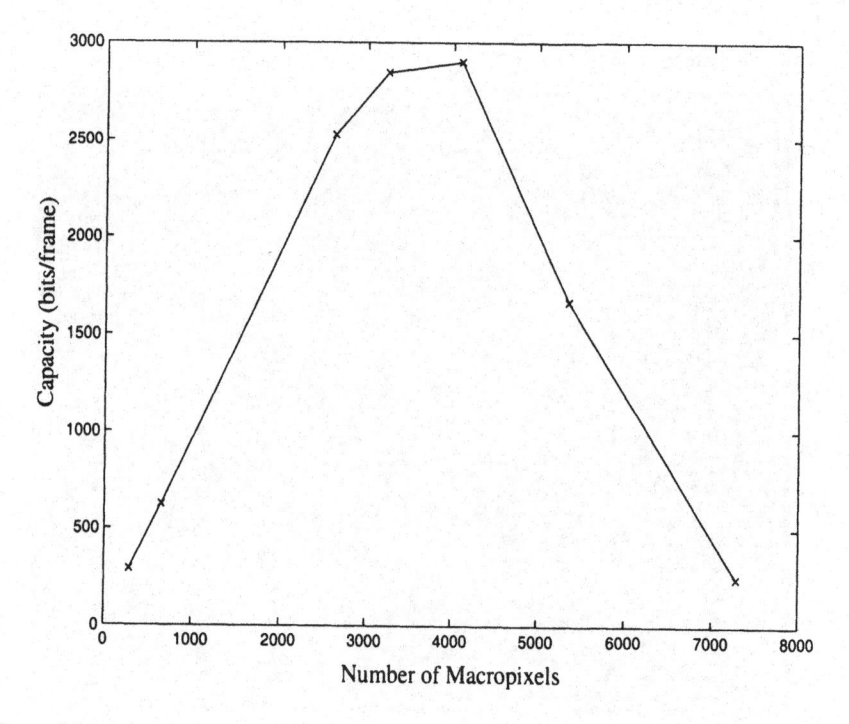

Figure 8.16. Approximate capacity in bits per frame versus the number of macropixels.

and load the spatial frequency bins appropriately. Section 8.4.3 illustrates an example of this power allocation for the system under consideration.

The transmitted image is formed by taking the inverse fast Fourier transform (IFFT) on the 512×512 pixel image. As discussed in Section 7.5, a *cyclic extension* is appended around the transmitted frame to ease equalization at the receiver. Figure 8.17 illustrates a typical SDMT symbol with cyclic extension.

At the receiver, the 154×154 pixel received image is sampled in time and in space and put in spatial frequency domain using a FFT. It is assumed that the temporal synchronization is exact and that the timing bars of Section 8.1.3 are used. Additionally, it is assumed that there is no temporal ISI present in the channel. Following spatial sampling an interpolation step may need to take place to aid in spatial synchronization, as discussed in Section 7.5.2. The frequency bins are equalized, decoded and unloaded.

8.4.2 Dynamic Range Compression

In DMT systems the output signal is the sum of a large collection of independent sinusoids. The amplitude distribution of these signals is nearly Gaussian and they exhibit large peak-to-average ratios [193]. Indeed, the SDMT sys-

Cyclic Extension (8 pixels) — Data Block (512 × 512 pixels)

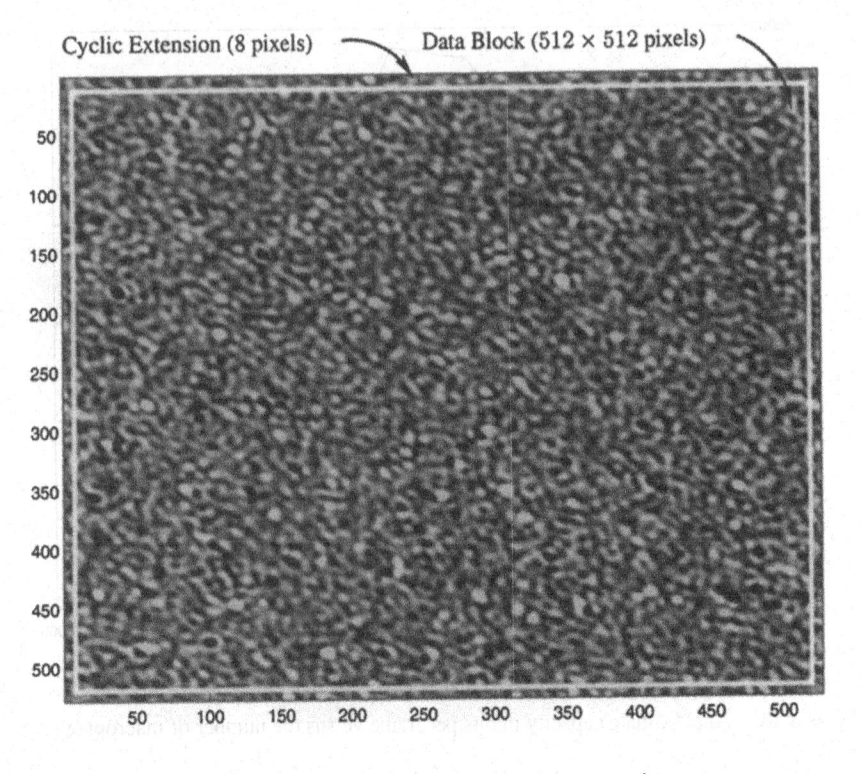

Figure 8.17.   An SDMT symbol with cyclic extension.

tem, being a generalization of conventional DMT, also exhibits this undesirable property.

In the case of SDMT, there are a large number of unused spatial frequency bins in which no data is set since the SNR is too low. These bins, however, present a degree of freedom and may be assigned to reduce the peak amplitude of the output signal. To assign these unused bins, an iterative projection technique [205] was employed. Figure 8.18 illustrates the operation of the algorithm. The algorithm starts by assigning the data symbols to the allocated spatial frequency bins and setting all other bins to zero. The resulting image is transformed to spatial domain using the IFFT, clipped to satisfy the peak constraint and placed back in spatial frequency domain using the FFT. The algorithm then assigns the data symbols to the frequency bins which are allocated and leaves all other bins untouched. This process iterates a fixed number of times. This algorithm can be viewed as an iterative projection between the set of signals satisfying the peak constraint and the set of signals carrying the required data. Since the two sets are convex, it can be shown that the algorithm will find a point in the intersection of the sets, if such a point exists [205].

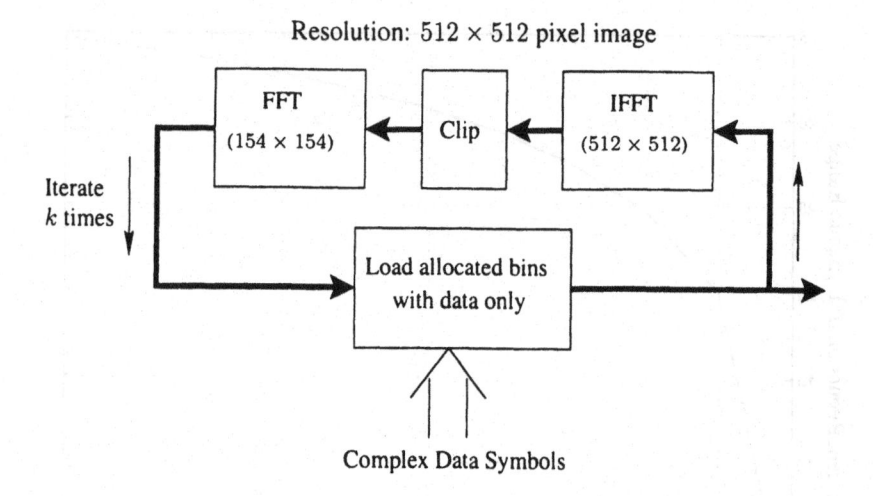

Figure 8.18. Block diagram of dynamic range compression algorithm.

Figure 8.19 shows the average reduction in the dynamic range versus the number of iterations when applied to SDMT with the power allocation found in Section 8.4.3. As the number of iterations increases, the dynamic range is also reduced at the cost of increased processing delay. For this link the algorithm was set to operate with five iterations where the reduction in dynamic range is nearly 30%.

By assigning values to unused bins, care must be taken that significant energy is not placed above the Nyquist band of the receiver which would add to aliasing noise. After five iterations, the resulting spectrum has less than 1% of the signal power outside of the Nyquist band. Note that these high spatial frequency bins will be additionally attenuated by the channel response before they are sampled at the receiver. This aliasing distortion impacts spatial frequency bins close to the Nyquist rate and results in a significant reduction in the dynamic range and the broad-band spatial clipping noise.

8.4.3 Signal Design and Capacity

The pixelated wireless optical channel considered here is a peak-limited channel. In general, it is difficult to use this constraint directly in frequency domain. It is clear, however, that imposing a limit on the maximum and minimum amplitudes restricts the electrical power which is transmitted per frame. This section computes an estimate of the capacity of the SDMT system on the channel model using the electrical water-pouring technique discussed in Section 7.5.3.

Figure 8.8 illustrates that ambient light produces a DC shift in the received intensities. As a result, the average amplitude of the image was set to $P = 117.5$

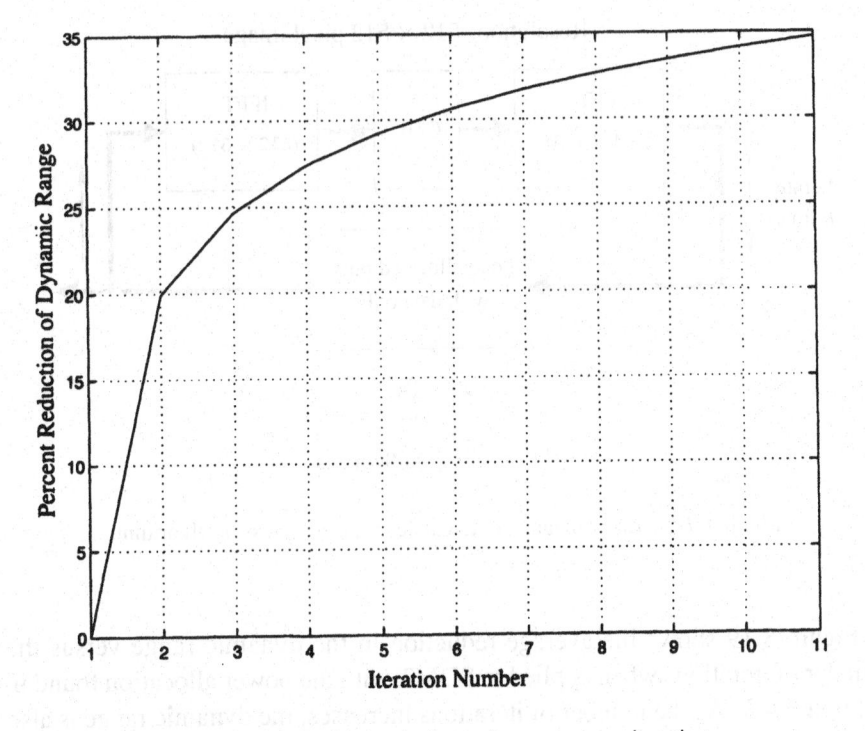

Figure 8.19.   Reduction in dynamic range versus iteration.

units. In order to arrive at a value for K in (7.11), each pixel was modelled as an independent random, Gaussian-distributed amplitude of mean P. Any amplitudes exceeding the range $[0, 255]$ are clipped by the transmitter. In order that the probability of clipping error was sufficiently small (less than 10^{-8}), the value of $K = 512^2 \times 0.03$ was taken.

The noise characteristics of the channel are signal-dependent, as characterized in Section 8.2.3. The noise is also composed of clipping and quantization noise. In this capacity estimate, the signal dependent characteristic of the noise sources are not considered directly. Rather, the noise spectrum per bin is estimated by running 50 frames through the system when the power allocation is set to spread E over all bins equally. The noise power received in each frequency bin within the Nyquist region of the receiver is then averaged to form an estimate of the noise spatial frequency distribution, $\sigma_n^2(u, v)$.

With an estimate of the noise variance in each bin, an electrical power allocation can be made amongst the spatial frequency bins in the Nyquist region of the receiver using the well-known water-pouring spectrum shown in (7.10) [45, 128]. Figure 8.20 presents a plot of the water-pouring "bowl" for the channel optical transfer function $H(u, v)$. The power allocation can be viewed as a process in which the constraint E is poured into the bowl and occupies those channels with the lowest $\sigma_n^2(u, v)/|H(u, v)|^2$. It should be noted that not all

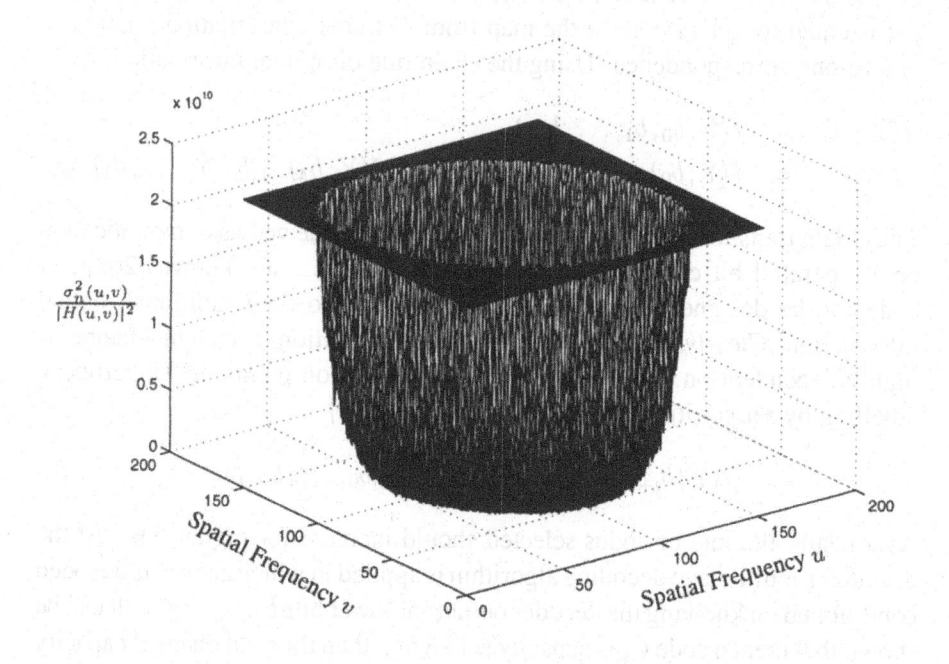

Figure 8.20. Water-pouring "bowl" for the SDMT channel.

spatial frequency points in the Nyquist band of receiver have power allocated to them. The vertical axis in Figure 8.20 is truncated to allow for plotting.

The capacity of the SDMT system can be estimated as the sum of a series of parallel Gaussian noise channels as in (7.9). Using the estimated noise variance per bin along with the computed power allocation the capacity of the channel is estimated to be 22.4 kbits/frame. This capacity does not explicitly take into account the peak constraint of the channel, but rather models the channel as an electrical channel with added Gaussian-distributed clipping noise.

8.4.4 Code Design

The capacity of the SDMT pixelated channel represents the maximum achievable rate per channel use (i.e, per frame), however, it does not suggest any realizable method to achieve it. In this section, a near-capacity achieving multi-level coding and multi-stage decoding scheme for DMT channels is applied to the SDMT channel [206].

Multi-level codes are a coded modulation scheme which use binary codes to improve the reliability of multi-level QAM constellations [207]. For constellations of size 2^M, M address bits, $B = (b_0, b_1, \ldots, b_{M-1})$, are required to label each constellation point. A multi-level coding scheme assigns a binary code, C_i, to each b_i depending on the "quality" of the given bit channel. Formally, if Y is the received variable, the mutual information for channel input to output is equal to $I(Y;B)$ since the map from B to the constellation points is a one-to-one correspondence. Using the chain rule of mutual information,

$$
\begin{aligned}
I(Y;B) &= I(Y;b_0,b_1,\ldots,b_M) \\
&= I(Y;b_0) + I(Y;b_1|b_0) + \cdots + I(Y;b_{M-1}|b_0,b_1,\ldots,b_{M-2}).
\end{aligned}
$$

Thus, data transmission on this channel can then be viewed as communication on M parallel bit channels, b_i, assuming $b_0,\ldots,b_{i-1}$ are known [208]. A code can be designed for each channel according to the conditional mutual information. Clearly, the conditional mutual information in each bit channel is highly dependent on the labelling of the constellation points. If Ungerboeck labelling by set partitioning [142] is used then [208]

$$
I(Y;b_{i+1}|b_0,\ldots,b_i) \geq I(Y;b_i|b_0,\ldots,b_{i-1}).
$$

As a result, the rate of codes selected should increase for higher bits. At the decoder, a multi-stage decoding algorithm is applied in which each b_i is decoded conditioned on knowing the decoder output for lower bits $b_0,\ldots,b_{i-1}$. It can be shown that if each code C_i is capacity achieving, then the total channel capacity is achieved using multi-level coding and multi-stage decoding [208].

In the context of DMT channels, it has been found that the use of multi-level codes can approach the channel capacity over a wide class of channels [206]. For the SDMT channel described in Section 8.4.3, multi-level codes are applied to approach the computed channel capacity.

Following [206], the bit loading algorithm selects the smallest constellation for each frequency bin so that the rate loss is at most 0.2 bits/symbol. The constellations are of size 2^i for $i = 1,\ldots,8$. Figures 8.21 and 8.22 present block diagrams of the multi-level coder and multi-stage decoder used for the SDMT channel. Bits b_0 and b_1 in each bin are Gray labelled and treated as a single symbol. Subject to the loading algorithm, the average capacity of b_0b_1 over all the frequency bins was computed to be 0.5563 bits/symbol. The intuition is that since there are a large number of frequency bins a long, powerful code with rate near the average capacity of b_0b_1 will perform well. These two bits in each spatial frequency bin are coded with a near capacity achieving, irregular rate-1/2 low density parity check code (LDPC) with block length 10^5 [206]. Simulations indicate that this code converges under the channel bin SNRs derived in Section 8.4.3. At the decoder, the log-likelihood ratios for each of these bits can be computed over all constellation points and fed to the LDPC decoder.

The higher level bits are labelled using Ungerboeck's set partitioning labelling where for each bit the intra-set distance increases. In order to design codes for the upper bits, a target bit-error rate of 10^{-7} was set. The upper bit

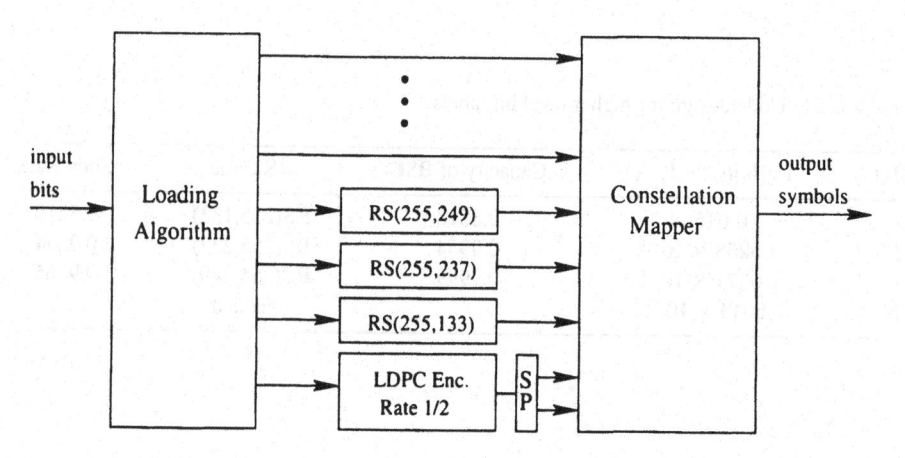

Figure 8.21. Multi-level coder block diagram for SDMT channel.

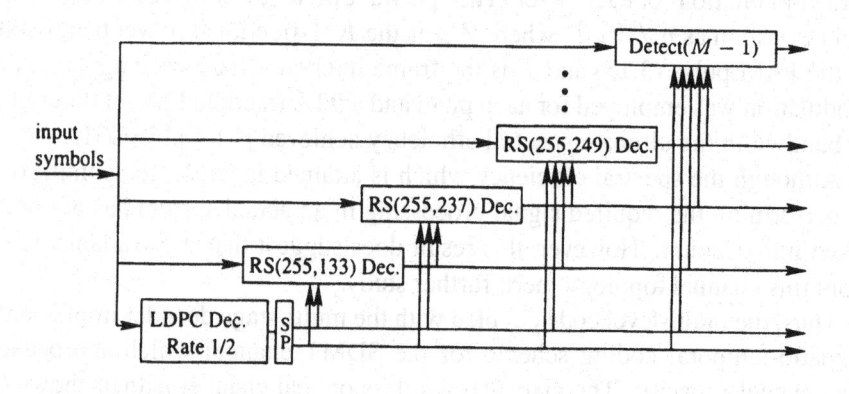

Figure 8.22. Multi-stage decoder for the SDMT channel.

channels are modelled as binary symmetric channels (BSCs) and hard decision Reed-Solomon codes of block length 255 were applied to correct enough errors to ensure that the target bit-error rate was met. Table 8.6 presents the average conditional probability of error for the higher order bits along with the capacity of the associated BSCs. For b_5 and higher label bits, the bit channel is good enough to allow for uncoded transmission while satisfying the bit-error rate target. For b_2, b_3 and b_4 the Reed-Solomon codes designed operate a rates close to the capacities of the binary symmetric channels.

After applying multi-level codes, the resulting rate over all frequency bins was computed to be $\eta_s = 17.1$ kbits/frame or approximately 76% of the estimated channel capacity. Normalized to the number of transmit and receive pixels, the transmission rate can also be stated as 0.061 bits/transmit pixel (including the cyclic extension) or 0.72 bits/receive pixel.

The spectral efficiency provides a metric of system performance which includes the temporal signalling constraints of the system. The spectral efficiency

Table 8.6.   Code design for higher level bit labels.

Bit	$Pe(b_i|b_0 \dots b_{i-1})$	Capacity of BSC	RS Code	Code Rate
b_2	0.01598	0.8818	RS(255,133)	0.5216
b_3	3.5258×10^{-4}	0.9954	RS(255,237)	0.9294
b_4	1.0371×10^{-6}	0.9999	RS(255,249)	0.9765
b_5	1.6014×10^{-11}	-	uncoded	-

of the system, η in (5.5), is dependent on the pulse shape used for the underlying PAM modulation for each pixel. The spectral efficiency of this SDMT scheme can be written as η_s/W_KT, where W_K is the $K\%$-fractional power bandwidth of the PAM pulse (3.15) and T is the frame interval. Using rectangular PAM modulation was employed for each pixel and a 99% fractional power definition of bandwidth the ultimate spectral efficiency achieved is 1.7 kbits/s/Hz.

Although the spectral efficiency which is attained is larger, the complexity of performing the required signal processing in an actual system has not been taken into account. However, this result does suggest that the available gains from this channel topology merit further study.

Thus, the multi-level code coupled with the multi-stage decoder implements a spatio-temporal coding scheme for the SDMT channel which approaches the channel capacity. The pixelated wireless optical channel is then shown to achieve high spectral efficiencies of approximately 1.7 kbits/s/Hz in a point-to-point configuration using a transmitter with 512×512 elements and a receiver with 154×154 elements over a range of 2 m.

8.5   Conclusions

The pixelated wireless optical channel is capable of providing large gains in spectral efficiency by leveraging the spatial diversity inherent to arrays of optical emitters and detectors. A channel model was developed based on channel measurements on a prototype channel. In simulation using the model, pixel-matched systems can achieve data rates of approximately 3 kbits/frame and does not consider the spatial frequency response of the channel. As a result, this technique suffer from severe ISI and is sensitive to spatial synchronization. This chapter discusses SDMT spatial modulation to combat the low pass spatial channel. A dynamic range compression technique was applied to exploit unused spatial frequency bins to reduce the peak value of output signals, thereby reducing clipping noise. A near-capacity achieving coding scheme, originally conceived for DMT systems, was applied to the SDMT channel. This spatio-temporal code achieves rates of approximately 17.1 kbits/frame in simulation which is 76% of the estimated channel capacity.

Although spatially invariant channels are discussed here, SDMT may also be appropriate on channels with some spatial variation. Consider the case of two-dimensional arrays of lasers and photodiodes for chip-to-chip signalling. Often deterministic variation of the gain of these devices over the array is present due to manufacturing errors [155]. Indeed, certain pixels may be totally inoperative. In conventional systems, where each transmit pixel is sensed by a single receive pixel, this distortion can cause erasures in the received data. If the spatial variation of these gains was know *a priori* at the transmitter, say during a calibration stage in manufacture, they could be taken into account in the power allocation algorithm. Thus, SDMT provides a degree of spatial redundancy which can accommodate a systematic spatial variation over the transmitter array.

In this chapter, the design of modulation and coding for a given pixelated wireless optical channel is considered. For systems in which the orthographic projection assumption holds, it is expected that the capacity of the system will be highly dependent on the range between transmitter and receiver. As the range increases, the spatial frequency response of the link drops. This assertion can be justified by assuming that a given image is transmitted and the range is increased. With reference to the receive imager, the sampling rate remains the same and the size of the received image scales down with increasing range implying a relative increase in the spatial frequency of the received image. However, as range increase a smaller number of receive pixels are illuminated. With a CMOS imager, the frame rate can be increased as the number of illuminated pixels reduces [182]. Thus, although the number of receiver pixels has decreased the signalling rate has improved. The trade-off between spatial and temporal bandwidth in this link is currently not well understood. Additionally, this work has not considered the impact of projective distortion on the capacity of the link.

Chapter 9

CONCLUSIONS AND FUTURE DIRECTIONS

9.1   Conclusions

This work has presented techniques for improving the spectral efficiency of wireless optical links. Existing signal design methods for electrical channels are extended and generalized to the optical intensity channel. Unlike previous wireless optical topologies, spatial diversity is exploited to improve the spectral efficiency of the channel.

A signal-space framework has been proposed which geometrically represents both amplitude constraints as well as the average optical power constraint. An additional peak constraint is also considered and represented in the signal space. Unlike previous techniques, this model is not confined to a given set of pulse shapes but treats all time disjoint signalling schemes in a common framework.

Lattice codes are defined using the signal space model. Conventional lattice coding techniques are extended to include the amplitude and average cost constraints of the optical intensity channel. A bandwidth constraint is imposed which represents the impact of shaping on the spectral characteristics of the modulation scheme. An idealized point-to-point link is used to compare the range over which modulation schemes can transmit subject to a constraint of bit-error rate and channel bandwidth. Spectrally efficient modulation is shown to provide significant rate increase for short distance channels.

In order to show the necessity of spectrally efficient modulation at high SNR, capacity bounds were computed for given pulse sets by exploiting the geometry of the signal space. The upper and lower bounds are shown to asymptotically exact at high optical signal-to-noise ratios. The bounds indicate that conventional rectangular PAM schemes used in optical channels suffer a large rate loss over bandwidth efficient techniques.

The MIMO wireless optical channel is defined as a general case of current angle diversity and quasi-diffuse channel structures. The spatial response of the channel is considered directly in the pixelated optical case and gains in spectral efficiency are had by transmitting information in both space and time. A prototype link is constructed using a commercial CCD camera and an LCD panel. Channel measurements are taken to characterize the channel impairments and a simulation model is defined for design work.

Spatial discrete multitone modulation is defined as a spatial signalling scheme which can adapt to the spatial bandwidth limitation of the channel. Spatial synchronization in this case is also shown to be easier than in the case of pixel-matched systems. A simulation study is performed to estimate the capacity of the channel and to propose coding to approach this capacity. Multi-level codes coupled with multi-stage decoders are applied to this channel and are shown to achieve 76% of the theoretical channel capacity and provide spectral efficiencies of approximately 1.7 kbits/s/Hz.

9.2   Future Work

In order to improve the range of wireless optical links the physical components of the link can be optimized in a number of ways. Optical concentrators, such as mirrors and lenses, can be used to increase the receive power at the price of higher implementation cost [2, 3]. Recently, holographic mirrors have been shown as a promising concentrator architecture [72]. At longer wavelengths (in the 1.3 μm and 1.5 μm range) the human eye is nearly opaque. As a result, an order of magnitude increase in the optical power transmitted can be realized at the price of costlier optoelectronics [11]. Using such techniques, the range over which high data rate spectrally efficient modulation schemes are appropriate can be extended at the expense of greater implementation cost.

Through the use of such techniques it is possible to engineer an optical channel which offers a significantly improved optical power at the receiver. Bandwidth efficient raised-QAM type modulation can then be applied in this new channel to provide improved data rates over the given transmission distance. Thus, these physical techniques increase the range of transmission distances in which high-rate, bandwidth efficient modulation is appropriate.

The signal space model of Chapter 4 represents only time-disjoint signals and their average cost. Since the channel is bandwidth constrained, the use of bandlimited pulses should be considered for this channel. A geometric representation of the amplitude constraint in the case of time-overlapping pulses is necessary to aid in signal design for this channel. Although some time-disjoint schemes are considered here, difficulty arises when considering high dimensional schemes due to the requirement that the cross-section Υ_1 be known. Bounds on the volume of Υ_1 for a given pulse set should be developed to allow the gain to be estimated for these cases.

Coding and shaping gain of lattice codes in Chapter 5 differs from the conventional results for electrical channels. The optical coding gain is not necessarily maximized by the densest lattice. It would be interesting to determine if the optical coding gain has an interpretation as the packing of some other geometric form. The lattice maximizing this packing is also an open topic. The use of opportunistic side channels is also proposed in the Chapter 5. It would be interesting to see if practical side channels can be constructed to aid in timing synchronization or to transmit additional information. The noise model of the wireless optical channel is also simplified and holds only at high background intensities. At lower background intensities the noise characteristics are certainly signal dependent. This situation may arise in chip-to-chip links where ambient light can be occluded as well as in narrow field-of-view links such as some MIMO wireless optical channels of Chapter 7. The design of signalling, and specifically lattice codes, subject to this distortion is also a promising avenue of future work.

The capacity of optical intensity signalling sets remains an open topic for investigation. Although asymptotically exact bounds are given here for a *given basis*, these results are not directly applicable in the synthesis of pulse sets. The basis set which maximizes the channel capacity subject to bandwidth and amplitude constraints still remains an open question. Studies into the capacity of MIMO channels corrupted by spatial distortions and signal dependent noise also show promise for future investigation.

The study of MIMO wireless optical channels presented in Chapters 7 and 8 present only an introduction to this area. Significant work still needs to be done in order to characterize these signalling schemes in the presence of more realistic distortion. However, this work has demonstrated that MIMO techniques, when applied to wireless optical channels, can yield impressive gains in spectral efficiency.

The MIMO wireless optical channel proposed in Chapter 7 can be realized in a wide range of configurations. The demonstration of this link using a larger number of transmit and receive elements and projection systems is of interest for future work. Non-planar flexible displays are also a potential implementation which would couple both the pixelated concept with a degree of angular diversity. The simulation model constructed is appropriate for the given fixed point-to-point link proposed. More general channel models should be formulated under a variety of channel configurations. The role of projective distortion should also be included in a more comprehensive channel model. Additionally, the spatial variation of the point-spread function and of noise should be more accurately modelled.

Spatial discrete multitone, investigated in simulation in Chapter 8, is shown to be a powerful extension of conventional DMT systems. Future work should include implementing this modulation scheme on a channel to verify the sim-

ulated performance. Additional work is also required to develop spatial registration and spatial synchronization algorithms which are critically important to
the functioning of this link. Lower complexity spatial modulation should also
be considered for this link. Two-dimensional partial response techniques [166]
should be studied in greater detail for MIMO wireless optical channels and the
available spectral efficiencies determined.

References

[1] F. R. Gfeller and U. Bapst. Wireless in-house communication via diffuse infrared radiation. *Proceedings of the IEEE*, 67(11):1474–1486, November 1979.

[2] J. R. Barry. *Wireless Infrared Communications*. Kluwer Academic Publishers, Boston, MA, 1994.

[3] R. Otte , L. P. de Jong and A. H. M. van Roermund. *Low-Power Wireless Infrared Communications*. Kluwer Academic Publishers, Boston, MA, 1999.

[4] S. Hranilovic. *Spectrally Efficient Signalling for Wireless Optical Intensity Channels*. PhD thesis, Dept. of Elec. & Comp. Engineering, University of Toronto, 2003.

[5] S. Hranilovic. Modulation and constrained coding techniques for wireless infrared communication channels. Master's thesis, Dept. of Elec. & Comp. Engineering, University of Toronto, 1999.

[6] Institute of Electrical and Electronics Engineers. Standards Association Website. standards.ieee.org.

[7] Infrared Data Association. Web Site. www.irda.org.

[8] T. Standage. *The Victorian Internet: the remarkable story of the telegraph and the nineteenth century's on-line pioneers*. Walker and Co., New York, NY, 1998.

[9] A. G. Bell. Selenium and the photophone. *Nature*, pages 500–503, Sept. 23, 1880.

[10] R. C. Weast, editor. *CRC Handbook of Chemistry and Physics*. CRC Press, Boca Raton, FL, 64th edition, 1983.

[11] J. M. Kahn and J. R. Barry. Wireless infrared communications. *Proceedings of the IEEE*, 85(2):263–298, February 1997.

[12] International Electrotechnical Commission. Safety of laser products – Part 1: Equipment classification, requirements and user's guide. Group safety publication, reference number 825-1, 1993.

[13] S. Bloom, E. Korevaar, J. Schuster, and H. Willebrand. Understanding the performance of free-space optics. *OSA Journal of Optical Networking*, 2(6):178–200, June 2003.

[14] P. L. Eardley and D. R. Wisely. 1 Gbit/s optical free space link operating over 40m – system and applications. *IEE Proceedings on Optoelectronics*, 143(6):330–333, December 1996.

[15] D. J. T. Heatley, D. R. Wisely, I. Neild, and P. Cochrane. Optical wireless: The story so far. *IEEE Communications Magazine*, pages 72–82, December 1998.

[16] Y. Tanaka, S. Haruyama, and M. Makagawa. Wireless optical transmissions with white colored LED for wireless home links. In *Proceedings of the IEEE International Symposium on Personal, Indoor and Mobile Radio Communication*, volume 2, pages 1325–1329, 2000.

[17] Talking Lights LLC. Web Site. www.talking-lights.com.

[18] D. R. Goff. *Fiber Optic Reference Guide — A Practical Guide to the Technology*. Focal Press, Boston, MA, 1996.

[19] S. Haykin. *Digital Communications*. John Wiley & Sons, New York, NY, 1988.

[20] J. M. Kahn, W. J. Krause, and J. B. Carruthers. Experimental characterization of non-directed indoor infrared channels. *IEEE Transactions on Communications*, 43(2/3/4):1613–1623, Feb./Mar./Apr. 1995.

[21] J. B. Carruthers and J. M. Kahn. Modeling of nondirected wireless infrared channels. *IEEE Transactions on Communications*, 45(10):1260–1268, October 1997.

[22] J. R. Barry, J. M. Kahn, W. J. Krause, E. A. Lee, and D. G. Messerschmitt. Simulation of multipath impulse response for indoor wireless optical channels. *IEEE Journal on Selected Areas in Communications*, 11(3):367–379, April 1993.

[23] G. W. Marsh and J. M. Kahn. Performance evaluation of experimental 50-Mb/s diffuse infrared wireless link using on-off keying with decision-feedback equalization. *IEEE Transactions on Communications*, 44(11):1496–1504, November 1996.

[24] V. Jungnickel, V. Pohl, S. Nönnig, and C. von Helmont. A physical model of the wireless infrared communication channel. *IEEE Journal on Selected Areas in Communications*, 20(3):631–640, April 2002.

[25] Y. A. Alqudah and M. Kavehrad. MIMO characterization of indoor wireless optical link using a diffuse-transmission configuration. *IEEE Transactions on Communications*, 51(9):1554–1560, September 2003.

[26] H. Hashemi, G. Yun, M. Kavehrad, F. Behbahani, and P. A. Galko. Indoor propagation measurements at infrared frequencies for wireless local area networks applications. *IEEE Transactions on Vehicular Technology*, 43(3):562–576, August 1994.

[27] M. R. Pakravan, M. Kavehrad, and H. Hashemi. Indoor wireless infrared channel characterization by measurements. *IEEE Transactions on Vehicular Technology*, 50(4):1053–1073, July 2001.

[28] W. B. Leigh. *Devices for Optoelectronics*. Marcel Dekker, Inc., New York, NY, 1996.

[29] M. Fukuda. *Reliability and Degradation of Semiconductor Lasers and LEDs*. Artech House, Boston, MA, 1991.

[30] D. H. Navon. *Semiconductor Microdevices and Materials*. Holt, Rinehart and Winston, New York, NY, 1986.

[31] Siemens Microelectronics. General IR and photodetector information. Application note 37, www.infineon.com/products/37/37.html, 1997.

[32] J. Straus. Linearized transmitters for analog fiber links. *Laser Focus*, pages 54–61, October 1978.

[33] D. Hassin and R. Vahldieck. Feedforward linearization of analog modulated laser diodes–theoretcial analysis and experimental verification. *IEEE Transactions on Microwave Theory and Techniques*, 41(12):2376–2382, December 1993.

[34] M. Bertelsmeier and W. Zschunke. Linearization of broadband optical transmission systems by adaptive predistortion. *Frequenz*, 38(9):206–212, 1984.

[35] H. Kressel, M. Ettenberg, J. P. Wittke, and I. Ladany. Laser diodes and LEDs for fiber optical communication. In H. Kressel, editor, *Semiconductor Devices for Optical Communication*, chapter 2, pages 9–62. Springer-Verlag, Berlin, Germany, 1980.

[36] R. P. Feynman, R. B. Leighton, and M. Sands. *The Feynman Lectures on Physics : Vol I*. Addson-Wesley Publishing, Reading, MA, 1977.

[37] Analog Devices Inc. AD9660 laser diode driver with light power control. Data sheet, One Technology Way, Norwood, MA, 1995.

[38] P. W. Shumate and M. DiDomenico Jr. Lightwave transmitters. In H. Kressel, editor, *Semiconductor Devices for Optical Communication*, chapter 5, pages 161–200. Springer-Verlag, Berlin, Germany, 1980.

[39] H. Melchior, M. B. Fisher, and F. R. Arams. Photodetectors for optical communication systems. *Proceedings of the IEEE*, 58(10):1466–1486, October 1970.

[40] S. B. Alexander. *Optical Communication Receiver Design*. Institution of Electrical Engineers, SPIE Optical Engineering Press, London, UK, 1997.

[41] D. P. Schinke, R. G. Smith, and A. R. Hartman. Photodetectors. In H. Kressel, editor, *Semiconductor Devices for Optical Communication*, chapter 3, pages 61–87. Springer-Verlag, Berlin, Germany, 1980.

[42] R. R. Hayes and D. L. Persechini. Nonlinearity of p-i-n photodetectors. *IEEE Photonics Technology Letters*, 5(1):70–72, January 1993.

[43] S. Karp, E. L. O'Neill, and R. M. Gagliardi. Communication theory for the free-space optical channel. *Proceedings of the IEEE*, 58(10):1611–1626, October 1970.

[44] R. G. Smith and S. D. Personick. Receiver design for optical fiber communication systems. In H. Kressel, editor, *Semiconductor Devices for Optical Communication*, chapter 4, pages 89–160. Springer-Verlag, Berlin, Germany, 1980.

[45] E. A. Lee and D. G. Messerschmitt. *Digital Communication*. Kluwer Academic Publishers, Boston, MA, 2nd edition, 1994.

[46] P. P. Webb, R. J. McIntyre, and J. Conradi. Properties of avalanche photodiodes. *RCA Review*, 35:234–278, June 1974.

[47] P. Balaban. Statistical evaluation of the error rate of the fiberguide repeater using importance sampling. *The Bell System Technical Journal*, 55(6):745–766, July-August 1976.

[48] G. P. Agrawal. *Fiber-Optic Communication Systems*. John Wiley and Sons, Toronto, Canada, 3rd edition, 2002.

[49] R. Narasimhan, M. D. Audeh, and J. M. Kahn. Effect of electronic-balast fluorescent lighting on wireless infrared links. *IEE Proceedings on Optoelectronics*, 143 6:347–354, December 1996.

[50] K. Phang and D. A. Johns. A CMOS optical preamplifier for wireless infrared communications. *IEEE Transactions on Circuits and Systems – II: Analog and Digital Signal Processing*, 46(7):852–859, July 1999.

[51] B. Zand, K. Phang, and D. A. Johns. Transimpedance amplifier with differential photodiode current sensing. In *Proceedings of the IEEE International Symposium on Circuits and Systems*, volume II, pages 624–627, 1999.

[52] Infrared Data Association. Infrared Data Association serial infrared physical layer specification. Version 1.4, www.irda.org, 2001.

[53] T. S. Chu and M. J. Gans. High speed infrared local wireless communication. *IEEE Communications Magazine*, 25(8):4–10, August 1987.

[54] D. R. Wisely. A 1 Gbit/s optical wireless tracked architecture for ATM delivery. In *IEE Colloquium on Optical Free-Space Communication Links*, pages 1411–1417, 1996.

[55] A. M. Street, K. Samaras, D. C. O'Brien, and D. J. Edwards. High speed wireless IR-LANs using spatial addressing. In *Proceedings of the IEEE International Symposium on Personal, Indoor and Mobile Radio Communication*, pages 969–973, 1997.

[56] J. Bellon, M. J. N. Sibley, D. R. Wisely, and S. D. Greaves. Hub architecture for infrared wireless networks in office environments. *IEE Proceedings on Optoelectronics*, 146(2):78–82, April 1999.

[57] V. Jungnickel, A. Forck, T. Haustein, U. Krüger, V. Pohl, and C. von Helmolt. Electronic tracking for wireless infrared communications. *IEEE Transactions on Wireless Communications*, 2(5):989–999, September 2003.

[58] Harex InfoTech. Web Site. www.mzoop.com.

[59] Ingenico. Web Site. www.ingenico.com.

[60] JVC, VIPSLAN-10. Web Site. www.jvc-victor.co.jp.

[61] Plaintree Systems Inc. Web Site. www.plaintree.com.

[62] G. Tourgee G. Nykolak, P. R. Szajowski and H. Presby. 2.5 Gbit/s free space optical link over 4.4km. *IEE Electronics Letters*, 35(7):578–579, April 1st, 1999.

[63] K. Wilson and M. Enoch. Optical communications for deep space missions. *IEEE Communications Magazine*, 38(8):134–139, August 2000.

[64] Optical SETI Program. Web Site. seti.ucolick.org/optical.

[65] F. Gfeller and W. Hirt. Advanced infrared (AIr): Physical layer for reliable transmission and medium access. In *Proceedings of the IEEE International Zurich Seminar on Broadband Communications*, pages 77–84, 2000.

[66] R. R. Valadas, A. R. Tavares, A. M. de Oliveira Duarte, A. C. Moreira, and C. T. Lomba. The infrared physical layer of the IEEE 802.11 standard for wireless local area networks. *IEEE Communications Magazine*, 36(12):107–112, December 1998.

[67] Infra-Com. Web Site. www.infra-com.com.

[68] R. Want, B. N. Schilit, N. I. Adams, R. Gold, K. Petersen, D. Goldberg, J. R. Ellis, and M. Weiser. An overview of the PARCTAB ubiquitous computing experiment. *IEEE Personal Communications*, 2(6):28–43, December 1995.

[69] R. Want and A. Hopper. Active badges and personal interactive computing objects. *IEEE Transactions on Consumer Electronics*, 38(1):10–20, February 1992.

[70] M. T. Smith. Smart cards: Integrating for portable complexity. *IEEE Computer*, 31(8):110–115, August 1998.

[71] G. Yun and M. Kavehrad. Spot-diffusing and fly-eye receivers for indoor infrared wireless coummunications. In *Proceedings of the IEEE International Conference on Selected Topics in Wireless Communications*, pages 262–265, 1992.

[72] S. Jivkova and M. Kavehrad. Receiver designs and channel characterization for multi-spot high-bit-rate wireless infrared communications. *IEEE Transactions on Communications*, 49(12):2145–2153, December 2001.

[73] J. B. Carruthers and J. M. Kahn. Angle diversity for nondirected wireless infrared communication. *IEEE Transactions on Communications*, 48(6):960–969, June 2000.

[74] M. Kavehrad and S. Jivkova. Indoor broadband optical wireless communications: Optical subsystems designs and their impact on channel characteristics. *IEEE Wireless Communications*, 10(2):30–35, April 2003.

[75] ABB HAFO. 1A301 — high performance LED. Data sheet, Mitel Microelectronics, 1994.

[76] Temic. BPV 10 NF — high speed silicon pin photodiode. Data sheet, Telefunken Semiconductors, 1994.

[77] H. Park and J. R. Barry. Modulation analysis for wireless infrared communications. In *Proceedings of the IEEE International Conference on Communications*, pages 1182–1186, 1995.

[78] S. G. Wilson. *Digital Modulation and Coding*. Prentice-Hall, Upper Saddle River, NJ, 1996.

[79] D. Shiu and J. M. Kahn. Shaping and nonequiprobable signalling for intensity-modulated signals. *IEEE Transactions on Information Theory*, 45(7):2661–2668, November 1999.

[80] H. J. Landau and H. O. Pollak. Prolate spheroidal wave functions, Fourier analysis and uncertainty – III: The dimension of the space of essentially time- and band-limited signals. *The Bell System Technical Journal*, 41:1295–1336, July 1962.

[81] D. Slepian. On bandwidth. *Proceedings of the IEEE*, 64(3):292–300, March 1976.

[82] B. Slepian and H. O. Pollak. Prolate spheroidal wave functions, Fourier analysis and uncertainty – I. *The Bell System Technical Journal*, 40:43–63, January 1961.

[83] H. J. Landau and H. O. Pollak. Prolate spheroidal wave functions, Fourier analysis and uncertainty – II. *The Bell System Technical Journal*, 40:65–84, January 1961.

[84] J. G. Proakis. *Digital Communications*. McGraw-Hill, New York, NY, 1983.

[85] Infrared Data Association. Infrared Data Association serial infrared physical layer specification. Version 1.3, www.irda.org, 1998.

[86] J. B. Carruthers and J. M. Kahn. Multiple-subcarrier modulation for nondirected wireless infrared communication. *IEEE Journal on Selected Areas in Communications*, 14(3):538–546, April 1996.

[87] B. Reiffen and H. Sherman. An optimum demodulator for Poisson processes: Photon source detectors. *Proceedings of the IEEE*, 51:1316–1320, October 1963.

[88] K. Abend. Optimum photon detection. *IEEE Transactions on Information Theory*, 12(1):64–65, January 1966.

[89] R. M. Gagliardi and S. Karp. M-ary Poisson detection and optical communications. *IEEE Transactions on Communication Technology*, COM-17(2):208–216, April 1969.

[90] G. M. Lee and G. W. Schroeder. Optical pulse position modulation with multiple positions per pulsewidth. *IEEE Transactions on Communications*, 25:360–364, March 1977.

[91] H. Suiyama and K. Nosu. MPPM: A method for improving the band-utilization efficiency in optical PPM. *IEEE Journal of Lightwave Technology*, 7(3):465–472, March 1989.

[92] C. N. Georghiades. Modulation and coding for throughput-efficient optical systems. *IEEE Transactions on Information Theory*, 40(5):1313–1326, September 1994.

[93] D. Shiu and J. M. Kahn. Differential pulse-postion modulation for power-efficient optical communication. *IEEE Transactions on Communications*, 47(8):1201–1210, August 1999.

[94] Z. Ghassemlooy, A. R. Hayes, N. L. Seed, and E. D. Kaluarachchi. Digital pulse interval modulation for optical communications. *IEEE Communications Magazine*, 36(12):95–99, December 1998.

[95] T. Lüftner, C. Kröpl, R. Hagelauer, M. Huemer, R. Weigel, and J. Hausner. Wireless infrared communications with edge position modulation for mobile devices. *IEEE Wireless Communications*, 10(2):15–21, April 2003.

[96] D. C. M. Lee and J. M. Kahn. Coding and equalization for PPM on wireless infrared channels. *IEEE Transactions on Communications*, 47(2):255–260, February 1999.

[97] M. D. Audeh, J. M. Kahn, and J. R. Barry. Decision-feedback equalization of pulse-position modulation on measured nondirected indoor infrared channels. *IEEE Transactions on Communications*, 47(4):500–503, February 1999.

[98] K. Kiasaleh. Turbo-coded optical ppm communication systems. *IEEE Journal of Lightwave Technology*, 16(1):18–26, January 1998.

[99] M. D. Audeh, J. M. Kahn, and J. R. Barry. Performance of pulse-position modulation on measured non-directed indoor infrared channels. *IEEE Transactions on Communications*, 44(6):654–659, June 1996.

[100] G. J. Pottie. Trellis codes for the optical direct-detection channel. *IEEE Transactions on Communications*, 39(8):1182–1183, August 1991.

[101] G. E. Atkin and K. L. Fung. Coded multipulse modulation in optical communication systems. *IEEE Transactions on Communications*, 42(2/3/4):574–582, Feb./Mar./Apr. 1994.

[102] H. Park and J. R. Barry. Trellis-coded multiple-pulse position modulation for wireless infrared communications. In *Proceedings of the IEEE Global Communications Conference*, volume 1, pages 225–230, 1998.

[103] R. M. Gagliardi and S. Karp. *Optical Communications*. John Wiley & Sons, New York, NY, 1976.

[104] S. Hranilovic and D. A. Johns. A multilevel modulation scheme for high-speed wireless infrared communications. In *Proceedings of the IEEE International Symposium on Circuits and Systems*, volume VI, pages 338–341, 1999.

[105] A. Garcia-Zambrana and A. Puerta-Notario. Improving PPM schemes in wireless infrared links at high bit rates. *IEEE Communications Letters*, 5(3):95–97, March 2001.

[106] R. You and J. M. Kahn. Average power reduction techniques for multiple-subcarrier intensity-modulated optical signals. *IEEE Transactions on Communications*, 49(12):2164–2171, December 2001.

[107] S. Walklin and J. Conradi. Multilevel signaling for increasing the reach of 10 Gb/s lightwave systems. *IEEE Journal of Lightwave Technology*, 17(11):2235–2248, November 1999.

[108] S. Hranilovic and F. R. Kschischang. Optical intensity-modulated direct detection channels: Signal space and lattice codes. *IEEE Transactions on Information Theory*, 49(6):1385–1399, June 2003.

[109] F. R. Kschischang and S. Pasupathy. Optimal nonuniform signalling for Gaussian channels. *IEEE Transactions on Information Theory*, 39(3):913–929, May 1993.

[110] A. Gray. *Modern Differential Geometry of Curves and Surfaces*. CRC Press, Boca Raton, FL, 1993.

[111] H. F. Harmuth. *Transmission of Information by Orthogonal Functions*. Springer-Verlag, New York, NY, 2nd edition, 1972.

[112] G. D. Forney Jr. and L.-F. Wei. Multidimensional constellations – Part I: Introduction, figures of merit, and generalized cross constellations. *IEEE Journal on Selected Areas in Communications*, 7(6):877–892, August 1989.

[113] R. J. McEliece. Practical codes for photon communication. *IEEE Transactions on Information Theory*, IT-27(4):393–398, July 1981.

[114] G. D. Forney Jr., R. G. Gallager, G. R. Lang, F. M Longstaff, and S. U. Qureshi. Efficient modulation for band-limited channels. *IEEE Journal on Selected Areas in Communications*, SAC-2(5):632–647, September 1984.

[115] G. D. Forney Jr. Coset codes — Part I: introduction and geometrical classification. *IEEE Transactions on Information Theory*, 34(5):1123–1151, September 1988.

[116] G. R. Lang and F. M. Longstaff. A Leech lattice modem. *IEEE Journal on Selected Areas in Communications*, 7(6):968–973, August 1989.

[117] A. R. Calderbank and L. H. Ozarow. Nonequiprobable signaling on the Gaussian channel. *IEEE Transactions on Information Theory*, 36(4):726–740, July 1990.

[118] F. R. Kschischang and S. Pasupathy. Optimal shaping properties of the truncated polydisc. *IEEE Transactions on Information Theory*, 40(3):892–903, May 1994.

[119] A. K. Khandani and P. Kabal. Shaping multidimensional signal spaces – Part I: optimum shaping, shell mapping. *IEEE Transactions on Information Theory*, 39(6):1799–1808, November 1993.

[120] J. H. Conway and N. J. A. Sloane. *Sphere Packings, Lattices and Groups*. Springer-Verlag, New York, NY, 2nd edition, 1993.

[121] G. D. Forney Jr. and G. Ungerboeck. Modulation and coding for linear Gaussian channels. *IEEE Transactions on Information Theory*, 44(6):2384–2415, October 1998.

[122] E. Biglieri and G. Caire. Power spectrum of block-coded modulation. *IEEE Transactions on Communications*, 42(2/3/4):1580–1585, Feb./Mar./Apr. 1994.

[123] Waterloo Maple Inc. *Maple V Programming Guide : Release 5*. Springer-Verlag, New York, NY, 1997.

[124] Hewlett Packard. Compliance of infrared communication products to IEC 825-1 and CENELEC EN 60825-1. Application note 1118, www.semiconductor.agilent.com/ir/app_index.html, 1997.

[125] K. L. Sterckx, J. M. Elmirghani, and R. A. Cryan. On the use of adaptive threshold detection in optical wireless communication systems. In *Proceedings of the IEEE Global Communications Conference*, volume 2, pages 1242–1246, 2000.

[126] S. Hranilovic and F. R. Kschischang. Capacity bounds for power- and band-limited optical intensity channels corrupted by Gaussian noise. *IEEE Transactions on Information Theory*, 50(5):784–795, May 2004.

[127] C. E. Shannon. A mathematical theory of communication. *The Bell System Technical Journal*, 27:379–423, 623–656, July, October 1948.

[128] T. M. Cover and J. A. Thomas. *Elements of Information Theory*. Wiley & Sons Publishers, New York, NY, 1991.

[129] R. G. Gallager. *Information Theory and Reliable Communication*. John Wiley & Sons Canada, Toronto, ON, 1968.

[130] R. W. Yeung. *A First Course in Information Theory*. Kluwer Academic Publishers, Boston, MA, 2002.

[131] J. P. Gordon. Quantum effects in communication systems. *Proceedings of the IRE,* pages 1898–1908, September 1962.

[132] J. R. Pierce. Optical channels: Practical limits with photon counting. *IEEE Transactions on Communications,* COM-26(12):1819–1821, December 1978.

[133] M. H. A. Davis. Capacity and cutoff rate for Poisson-type channels. *IEEE Transactions on Information Theory,* IT-26(6):710–715, November 1980.

[134] A. D. Wyner. Capacity and error exponent for the direct detection photon channel — Part I and II. *IEEE Transactions on Information Theory,* 34(6):1449–1471, November 1988.

[135] S. Shamai (Shitz). Capacity of a pulse amplitude modulated direct detection photon channel. *IEE Proceedings,* 137(6):424–430, December 1990.

[136] T. H. Chan, S. Hranilovic, and F. R. Kschischang. Capacity-achieving probability measure for conditional Gaussian channels with bounded inputs. under review *IEEE Trans. Inform. Theory.*

[137] H. Park and J. R. Barry. Performance analysis and channel capacity for multiple-pulse position modulation on multipath channels. In *Proceedings of the IEEE International Symposium on Personal, Indoor and Mobile Radio Communication,* volume 1, pages 247–251, 1996.

[138] R. You and J. M. Kahn. Upper-bounding the capacity of optical IM/DD channels with multiple-subcarrier modulation and fixed bias using trigonometric moment space method. *IEEE Transactions on Information Theory,* 48(2):514–523, February 2002.

[139] C. E. Shannon. Communication in the presence of noise. *Proceedings of the IRE,* 37(1):10–21, January 1949.

[140] J. M. Wozencraft and I. M. Jacobs. *Principles of Communication Engineering.* John Wiley & Sons, New York, NY, 1965.

[141] W. Feller. *An Introduction to Probability Theory and Its Applications,* volume 1. John Wiley and Sons Inc., New York, NY, 3rd edition, 1968.

[142] G. Ungerboeck. Channel coding with multilevel/phase signals. *IEEE Transactions on Information Theory,* IT-28(1):55–67, January 1982.

[143] Mathworks Inc. *MATLAB Version 5.3.* Prentice-Hall, Upper Saddle River, NJ, 1999.

[144] D. Slepian. Prolate spheroidal wave functions, Fourier analysis and uncertainty – V: The discrete case. *The Bell System Technical Journal,* 57:1371–1430, May-June 1978.

[145] S. Hranilovic and F. R. Kschischang. A short-range wireless optical channel using pixelated transmitters and imaging receivers. under review *IEEE Trans. Commun.*

[146] S. Hranilovic and F. R. Kschischang. Short-range wireless optical communication using pixelated transmitters and imaging receivers. In *Proceedings of the IEEE International Conference on Communications,* 2004.

[147] J. M. Kahn, R. You, P. Djahani, A. G. Weisbin, B. K. Teik, and A. Tang. Imaging diversity receivers for high-speed infrared wireless communication. *IEEE Communications Magazine*, 36(12):88–94, December 1998.

[148] P. Djahani and J. M. Kahn. Analysis of infrared wireless links employing multibeam transmitters and imaging diversity receivers. *IEEE Transactions on Communications*, 48(12):2077–2088, December 2000.

[149] K. Akhavan, M. Kavehrad, and S. Jivkova. High-speed power-efficient indoor wireless infrared communication using code combining – Part I. *IEEE Transactions on Communications*, 50(7):1098–1109, July 2002.

[150] K. Akhavan, M. Kavehrad, and S. Jivkova. High-speed power-efficient indoor wireless infrared communication using code combining – Part II. *IEEE Transactions on Communications*, 50(9):1495–1502, September 2002.

[151] S. M. Haas, J. H. Shapiro, and V. Tarokh. Space-time codes for wireless optical channels. In *Proceedings of the IEEE International Symposium on Information Theory*, page 244, Washington D.C., USA, June 2001.

[152] S. M. Haas, J. H. Shapiro, and V. Tarokh. Space-time codes for wireless optical communications. *EURASIP Journal on Applied Signal Processing*, 2002(3):211–220, March 2002.

[153] X. Zhu and J. M. Kahn. Free-space optical communication through atmospheric turbulence channels. *IEEE Transactions on Communications*, 50(8):1293–1300, August 2002.

[154] D. V. Plant, M. B. Venditti, E. Laprise, J. Faucher, K. Razavi, M. Châteauneuf, A. G. Kirk, and J. S. Ahearn. 256-channel bidirectional optical interconnect using VCSELs and photodiodes on cmos. *IEEE Journal of Lightwave Technology*, 19(8):1093–1103, August 2001.

[155] M. Châteauneuf, A. G. Kirk, D. V. Plant, T. Yamamoto, and J. D. Ahearn. 512-channel vertical-cavity surface-emitting laser based free-space optical link. *IEEE Journal of Lightwave Technology*, 41(26):5552–5561, September 2002.

[156] D. V. Plant and A. G. Kirk. Optical interconnects at the chip and board level: Challenges and solutions. *Proceedings of the IEEE*, 88(6):806–818, June 2000.

[157] M. A. Neifeld and R. K. Kostuk. Error correction for free-space optical interconnects: Space-time resource optimization. *Optics Letters*, 37(2):296–307, January 1998.

[158] T. H. Szymanski. Bandwidth optimization of optical data link by use of error-control codes. *Optics Letters*, 39(11):1761–1775, April 2000.

[159] J. Faucher, M. B. Venditti, and D. V. Plant. Application of parallel forward-error correction in two-dimensional optical-data links. *IEEE Journal of Lightwave Technology*, 21(2):466–475, February 2003.

[160] G. T. Sincerbox. History and physical principles. In H. J. Coufal, D. Psaltis, and G. T. Sincerbox, editors, *Holographic Data Storage*, chapter 1, pages 3–20. Springer-Verlag, Berlin, Germany, 2000.

[161] D. Psaltis and G. W. Burr. Holographic data storage. *IEEE Computer*, 312:52–60, February 1998.

[162] R. M. Shelby, J. A. Hoffnagle, G. W. Burr, C. M. Jefferson, M.-P. Bernal, H. Coufal, R. K. Grygier, H. Günther, R. M. Macfarlane, and G. T. Sincerbox. Pixel-matched holographic data storage with megabit pages. *Optics Letters*, 22(19):1509–1511, October 1997.

[163] G. W. Burr, J. Ashley, H. Coufal, R. K. Grygier, J. A. Hoffnagle, C. Michael Jefferson, and B. Marcus. Modulation coding for pixel-matched holographic data storage. *Optics Letters*, 22(9):639–641, May 1997.

[164] C. Gu, P. Yeh, X. Yi, and J. Hong. Fundamental noise sources in volume holographic storage. In H. J. Coufal, D. Psaltis, and G. T. Sincerbox, editors, *Holographic Data Storage*, chapter 3, pages 63–60. Springer-Verlag, Berlin, Germany, 2000.

[165] B. V. K. Vijaya Kumar, V. Vadde, and M. Keskinoz. Equalization for volume holographic data storage systems. In H. J. Coufal, D. Psaltis, and G. T. Sincerbox, editors, *Holographic Data Storage*, chapter 19, pages 309–318. Springer-Verlag, Berlin, Germany, 2000.

[166] B. H. Olson and S. C. Esener. One and two dimensional parallel partial response for parallel readout optical memories. In *Proceedings of the IEEE International Symposium on Information Theory*, page 141, 1995.

[167] G. W. Burr. Holographic data storage with arbitrarily misaligned data pages. *Optics Letters*, 27(7):542–544, April 2002.

[168] G. W. Burr and T. Weiss. Compensation for pixel misregistration in volume holographic data storage. *Optics Letters*, 26(8):542–544, April 2001.

[169] B. Marcus. Modulation codes for holographic recording. In H. J. Coufal, D. Psaltis, and G. T. Sincerbox, editors, *Holographic Data Storage*, chapter 17, pages 283–292. Springer-Verlag, Berlin, Germany, 2000.

[170] M. A. Neifeld and W.-C Chou. Interleaving and error correction for holographic storage. In H. J. Coufal, D. Psaltis, and G. T. Sincerbox, editors, *Holographic Data Storage*, chapter 18, pages 293–308. Springer-Verlag, Berlin, Germany, 2000.

[171] B. G. Boone. *Signal Processing Using Optics: Fundamentals, Devices, Architectures, and Applications*. Oxford University Press, New York, NY, 1998.

[172] Texas Instruments, Digital Light Processing. Web Site. www.dlp.com.

[173] Kodak Display Products. Web Site. www.kodak.com/US/en/corp/display/.

[174] Y. Chen, J. Au, P. Kazlas, A. Ritenour, H. Gates, and M. McCreary. Flexible active-matrix electronic ink display. *Nature*, 423:136, May 8, 2003.

[175] E. Bisaillon, D. F. Brosseau, T. Yamamoto, M. Mony, E. Bernier, D. Goodwill, D. V. Plant, and A. G. Kirk. Free-space optical link with spatial redundancy for misalignment tolerance. *IEEE Photonics Technology Letters*, 14(2):242–244, February 2002.

[176] H. F. Bare, F. Haas, D. A. Honey, D. Mikolas, H. G. Craighead, G. Pugh, and R. Soave. A simple surface-emitting LED array useful for developing free-space optical interconnects. *IEEE Photonics Technology Letters*, 5(2):172–174, February 1993.

[177] G. C. Holst. *CCD Arrays, Cameras, and Displays*. SPIE Optical Engineering Press, Bellingham, WA, 1996.

[178] Sony Electronics. CCD Catalog [available online]. www.sony.co.jp/ semicon/english/90203.html.

[179] B. S. Leibowitz, B. E. Boser, and K. S. J. Pister. CMOS "smart pixel" for free-space optical communication. *Proceedings SPIE*, 4306A:21–26, January 2001.

[180] D. C. O'Brien, G. E. Gaulkner, K. Jim, E. B. Zyamba, D. J. Edwards, M. Whitehead, P. Stavrinou, G. Parry, J. Bellon, M. J. Sibley, V. A. Lalithambika, V. M. Joyner, R. J. Samsudin, D. M. Holburn, and R. J. Mears. High-speed integrated transceivers for optical wireless. *IEEE Communications Magazine*, 41(3):58–62, March 2003.

[181] S. Kleinfelder, S. Lim, X. Liu, and A. El Gamal. A 10kframe/s $0.18\mu m$ CMOS digital pixel sensor with pixel-level memory. In *Proceedings of the IEEE International Solid-State Circuits Conference*, pages 88–89, 2001.

[182] A. El Gamal. EE 392B – Introduction to Image Sensors and Digital Cameras. Course notes [available online] www.stanford.edu/class/ee392b/, Stanford University, 2001.

[183] R. Klette, K. Schlüns, and A. Koschan. *Computer Vision: Three-Dimensional Data from Images*. Springer-Verlag, Singapore, 1998.

[184] Silicon Optix Inc. Web Site. www.siliconoptix.com.

[185] J. W. Goodman. *Introduction to Fourier Optics*. McGraw-Hill Book Company, New York, NY, 1968.

[186] W. F. Schreiber. *Fundamentals of Electronic Imaging Systems: Some Aspects of Image Processing*. Springer-Verlag, Berlin, Germany, 3rd edition, 1993.

[187] A. M. Bruckstein, L O'Gorman, and A. Orlitsky. Design of shapes for precise image registration. *IEEE Transactions on Information Theory*, 44(7):3156–3162, November 1998.

[188] C. B. Bose and I. Amir. Design of fiducials for accurate registration using machine vision. *IEEE Transactions on Pattern Analysis and Machine Intelligence*, 12(12):1196–1200, December 1990.

[189] G. J. Foschini and M. J. Gans. On limits of wireless communications in a fading environment when using multiple antennas. *Wireless Personal Communications*, 6(3):311–335, March 1998.

[190] I. E. Telatar. Capacity of multi-antenna Gaussian channels. *European Trans. Telecommun.*, 10(6):585–595, Nov./Dec. 1999.

[191] J. M. Cioffi. Asymmetric digital subscriber lines. In J. D. Gibson, editor, *Communications Handbook*, chapter 34, pages 450–479. CRC Press / IEEE Press, Boca Raton, FL, 1997.

[192] J. S. Chow, J. C. Tu, and J. M. Cioffi. A discrete multitone transceiver system for HDSL applications. *IEEE Journal on Selected Areas in Communications*, 9(6):895–908, August 1991.

[193] J. A. C. Bingham. Multicarrier modulation for data transmission: An idea whose time has come. *IEEE Communications Magazine*, pages 5–14, May 1990.

[194] R. C. Gonzalez and R. E. Woods. *Digital Image Processing*. Addison-Wesley, Reading, MA, 1992.

[195] H. Meyr and A. Polydoros. On sampling rate, analog prefiltering, and sufficient statistics for digital receivers. *IEEE Transactions on Communications*, 42(12):3208–3214, December 1994.

[196] A. F. Naguid, N. Sashadri, and A. R. Calderbank. Increasing data rate over wireless channels. *IEEE Signal Processing Magazine*, pages 76–92, May 2000.

[197] S. M. Alamouti. A simple transmit diversity technique for wireless communications. *IEEE Journal on Selected Areas in Communications*, 16(8):1451–1458, October 1998.

[198] G. D. Golden, C. J. Foschini, R. A. Valenzuela, and P. W. Wolniansky. Detection algorithm and initial laboratory results using V-BLAST space-time communication architecture. *IEE Electronics Letters*, 35(1):14–16, January 7th, 1999.

[199] NEC Solutions America. NEC Versa Web Site. support.neccomp.com.

[200] Basler Vision Technologies. Web Site. www.basler-vc.com.

[201] Tamron USA Inc. Web Site. www.tamron.com.

[202] Matrox Imaging. Web Site. www.matrox.com/imaging/.

[203] D. M. Etter, D. C. Kuncicky, and D. W. Hull. *Introduction to MATLAB 6.0*. Prentice-Hall, Upper Saddle River, NJ, 2001.

[204] N. R. Draper and H. Smith. *Applied Regression Analysis*. John Wiley & Sons, 2nd edition, 1981.

[205] A. Gatherer and M. Polley. Controlling clipping probability in DMT transmission. In *31st Asilomar Conference on Signals, Systems & Computers*, volume 1, pages 578–584, 1997.

[206] M. Ardakani, T. Esmailian, and F. R. Kschischang. Near-capacity coding in multi-carrier modulation systems. *IEEE Transactions on Communications*, submitted for publication.

[207] H. Imai and S. Hirakawa. A new multilevel coding method using error-correcting codes. *IEEE Transactions on Information Theory*, 23(3):371–377, May 1977.

[208] U. Wachsmann, R. F. H. Fischer, and J. B. Huber. Multilevel codes: Theoretical concepts and practical design rules. *IEEE Transactions on Information Theory*, 45(5):1361–1391, July 1999.

Index

About the Author

Steve Hranilovic is an Assistant Professor in the Department of Electrical and Computer Engineering, McMaster University, Hamilton, Ontario, Canada. He received the B.A.Sc. degree with honours in electrical engineering from the University of Waterloo, Canada in 1997 and M.A.Sc. and Ph.D. degrees in electrical engineering from the University of Toronto, Canada in 1999 and 2003 respectively. From 1992 to 1997, while studying for the B.A.Sc. degree, he worked in the areas of semiconductor device characterization and microelectronics for Nortel Networks and the VLSI Research Group, University of Waterloo. As a graduate student, Dr. Hranilovic received the Natural Sciences and Engineering Research Council of Canada postgraduate scholarships, Ontario Graduate Scholarship in Science and Technology, the Walter C. Sumner Foundation fellowship and the Shahid U. Qureshi Scholarship for Research in Communications. Dr. Hranilovic's research interests are in the areas of free-space and wired optical communications, digital communications algorithms, and electronic and photonic implementation of coding and communication algorithms.

From August 2000 until August 2003, Dr. Hranilovic served as co-Chair of the IEEE Communications Society, Toronto Chapter.